BALMAIN, GLEBE
and ANNANDALE
WALKS

Exploring the Suburbs

BALMAIN, GLEBE *and* ANNANDALE WALKS

Joan Lawrence

Hale & Iremonger

By the same author:
Sydney from Circular Quay
Sydney from The Rocks
North Shore Walks
Eastern Suburbs Walks (forthcoming)

© 1992 by Joan Lawrence

This book is copyright. Apart from any fair dealing for the purposes of study, research, criticism, review, or as otherwise permitted under the Copyright Act, no part may be reproduced by any process without written permission. Inquiries should be made to the publisher.

Typeset, printed & bound by
Southwood Press Pty Limited
80-92 Chapel Street, Marrickville, NSW

For the publisher
Hale & Iremonger Pty Limited
GPO Box 2552, Sydney, NSW

National Library of Australia Cataloguing-in-publication entry

Lawrence, Joan
    Balmain, Glebe and Annandale walks.

    Includes index.
    ISBN 0 86806 350 9.

    1. Walking — New South Wales — Sydney — Guidebooks. 2. Historic buildings — New South Wales — Sydney — Guidebooks. 3. Sydney (N.S.W.) — Description — 1990- — Guidebooks. I. Title.

919.4410463

Contents

	Illustrations	7
	Acknowledgements	9
Walk 1	Balmain	*11*
Walk 2	Balmain	*31*
Walk 3	Birchgrove	*43*
Walk 4	Annandale	*59*
Walk 5	Glebe	*81*
Profiles	Dr William Balmain	*103*
	Joseph Looke	*104*
	William Edmund Kemp	*104*
	William Morris Hughes	*105*
	Captain Thomas Stephenson Rowntree	*106*
	George Allen	*107*
	George Allen Mansfield	*107*
	Time Line	*109*
	Opening Times	*114*
	Further Reading	*115*
	Index	*117*

Illustrations

Timber house, Balmain   *13*

Nicholson Street Public School   *18*

Stone terrace, Peacock Point   *22*

Stone stairs to Union Street   *25*

Mort's Dock*   *28*

Balmain Watch House   *32*

John Booth, timber merchants*   *34*

Gladstone Park   *38*

Birchgrove Park and Snails Bay   *53*

Ravens Court   *54*

Annandale House*   *60*

Hunter Baillie Memorial Church   *65*

Daisy-patterned cast-iron   *66*

Beale Piano Factory   *68*

Edwinville   *69*

Uniting Church, Johnston Street, Annandale   *70*

'Witches' Houses'   *75*

Oybin   *75*

The Abbey   *76*
Glebe Point Road*   *82*
Chinese Joss House   *85*
St Scholastica's College   *87*
The Lodge   *90*
Record Reign Hall   *97*
Chesterfield House   *98*
Westmoreland Street   *99*

---

* Illustration courtesy Mitchell Library, State Library of New South Wales. All others are by the author.

Acknowledgements

Acknowledgement is made to Kath Hamey, guide, for planning the walks of Balmain and Birchgrove and for all her help.

Use has been made of the excellent Leichhardt Historical Journals and those interested in the Balmain/Birchgrove area should obtain copies of the publications by Peter Reynolds from the Architectural History Research Unit, Graduate School of the Built Environment, Faculty of Architecture, University of New South Wales. The publications are available at the Balmain Watch House, Darling Street, Balmain.

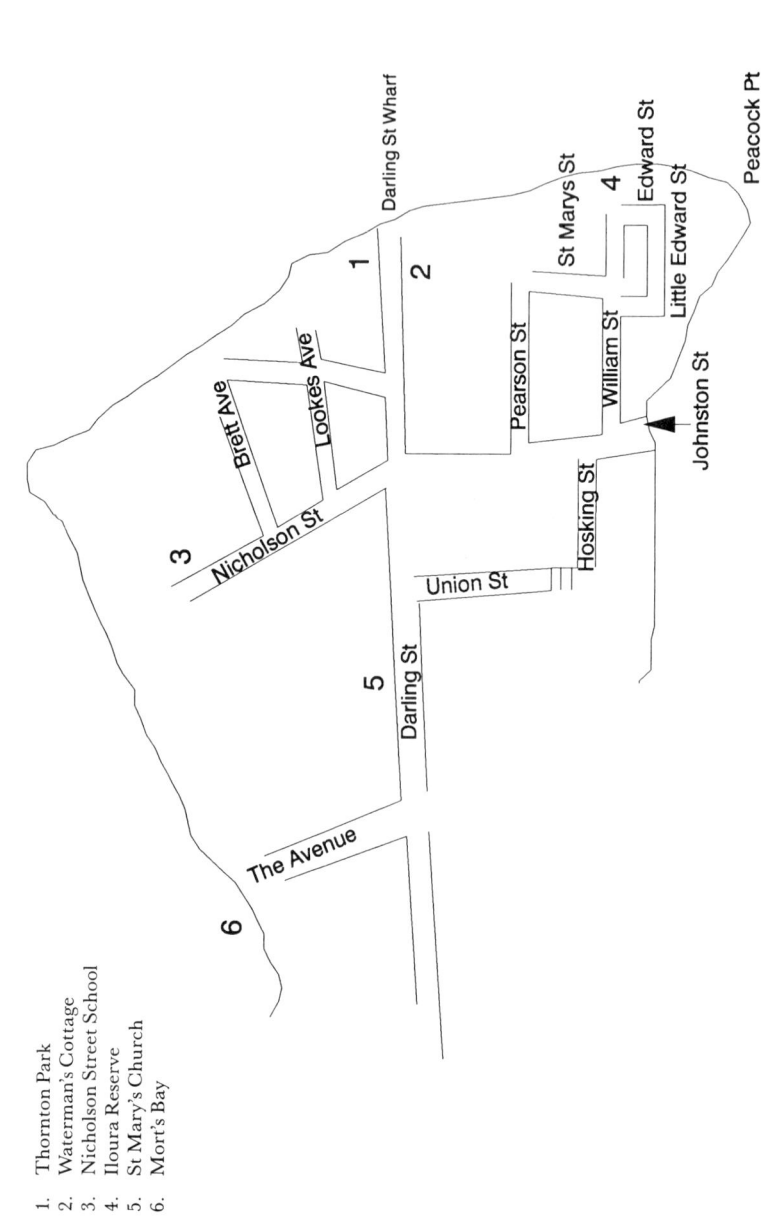

1. Thornton Park
2. Waterman's Cottage
3. Nicholson Street School
4. Iloura Reserve
5. St Mary's Church
6. Mort's Bay

Walk No. 1

Balmain

From Circular Quay to Darling Street, Peacock Point and Mort Bay

The most pleasant approach to Balmain is by ferry. Regular services operate from Circular Quay to the Darling Street wharf.

Balmain is a seafaring suburb: the deep waters of the harbour attracted the maritime industry. Captains settled on the point, building their homes close to the sailing vessels on the harbour. Balmain, the harbour and sailing are inseparable. The Balmain regattas were once part of the harbour scene and the suburb is credited with developing the 18-footers in the 1890s. They carried 2,800 square feet (260 sq. m.) of sail and crews of twelve to fifteen men to counterbalance the sails. The craft were derisively called 'troop ships'. They were so heavy they could travel at about only six knots. The 16-footers followed in 1901.

The first Balmain Regatta was held on 30 November 1849. Professional boat racing was a popular spectator sport in Sydney and the Balmain regattas encouraged rowing and sailing races. The races were extremely popular from the 1850s through to the 1870s. Many Sydney watermen participated in the races and there were large cash prizes to be won (*see* Walk 3). During Heritage Week the Balmain Regatta is briefly revived and people once more enjoy the spectacle on the harbour.

Balmain's post-1788 history starts with William Balmain, surgeon of the First Fleet (*see* Profile). On 26 April 1800 he was given a grant of 550 acres (223 ha.) on the peninsula west of Cockle Bay (Darling Harbour). Balmain named his grant Gilchrist Place and in July 1801 transferred it by 'bargain and sale' to John Gilchrist, head of Fort William College at Calcutta, India for 5 shillings. It seems possible the land transfer was security for a cargo of goods, including alcohol, from India. Gilchrist later settled in Sydney and became a merchant and shipowner. Between

1825 and 1831 he subdivided the grant and sold the first blocks at Peacock Point in October 1836 for £56 per acre (0.4 ha.).

The land was snapped up by the shipbuilding firms and Balmain's maritime history began. Dr Balmain's descendants refused to accept the legality of the transfer of the land and for more than eighty years they argued vainly for their 'lost' inheritance. The Gilchrist heirs also wrangled over 'unused land' for almost twenty years before the battle ended in the House of Lords, London. In 1865 the Gilchrist Educational Trust allowed the gradual sale of the Balmain land. The funds obtained were then used for education and libraries for working class people of the manufacturing towns in England. (For details of the land grant, see *Half a Thousand Acres: Balmain, a History of the Land Grant* Reynolds & Flottmann, 1976.)

The ten-minute ferry journey takes us past Sydney's historic Rocks and under the harbour bridge. To the left is Sydney's Wharf Theatre, an imaginative use of one of the old wharves, and on the right the bush-clad bulk of Ball's Head. The headland derives its name from Lieutenant Henry Lidgbird Ball, who was commander of the *Supply* in the First Fleet in 1788. Lieutenant Ball carried the first party of officers and convicts to Norfolk Island in February 1788 and discovered and named Lord Howe Island in honour of the British Admiral who had fought the North American colonists in the American War of Independence.

We soon pass Goat Island, called *Mel Mel* by the Aborigines, and gaze upon the Balmain Peninsula crowded with old houses, hilly streets, and patches of shrubbery and trees. The ferry glides into Darling Street wharf from which there are superb views of the city skyline and Observatory Hill. The tall Harbour Master's Tower watches over Sydney Harbour.

Walk commences at Darling Street Wharf

Balmain's maritime past is immediately apparent for there are old stone buildings on the left of the wharf. J. Fenwick & Co. Pty Ltd, tug owners, have been on this site for more than 100 years.

John and Thomas Fenwick began their tug service in 1870 with the twenty-six ton (26.5 tonnes) steam vessel, the *J. & T. Fenwick*. Thomas left the firm to go to the north coast where he based himself at Ballina and was involved in the coastal river trade. In 1883 John Fenwick bought the Darling Street land from the Bell family and he and his sons, James and Andrew, carried on their harbour business. John Fenwick died in 1901. Since that time alterations to the roof and arched doorway have been made to the old buildings.

The Fenwick tugs have been a familiar sight on Sydney Harbour. The *Castle Cove*, *Sydney Cove*, or *Manly Cove* would often be seen busily engaged on their work around the harbour wharves. The *Himma* was one of the last steam tugs in Sydney. It was built in England in 1942 and, before it

Old timber houses are typical of many parts of Balmain

came to Sydney, had served in the Persian Gulf during World War II. These days the tugs are diesel powered and Fenwicks have five tugs operating in Port Jackson and Port Botany.

There is another reminder of John Fenwick in Balmain. He once lived in St Mary's Street and his initials are in the leaded glass fanlight of his old front door.

Thornton Park was part of the land auctioned in 1836 and purchased by a Captain Adams. The land changed hands several tunes and at one time was owned by Peter Nichol Russell. He was born at Kirkcaldy in Scotland in 1816. His father, Robert Russell, was an engineer and ironfounder who migrated with his sons to Van Diemen's Land in 1832. He established the Hobart Town Foundry & Smithy and later transferred his business to Sydney. The sons were brass-founders, engineers and coppersmiths and had iron and machinery stores in Bridge Street, Sydney. Peter Russell had a foundry in George Street, and the firm supplied tools to the goldminers and the iron work for Victoria Barracks at Paddington and various gaols. Industrial unrest in the 1870s caused the closure of Russell's foundry. With wise investments Russell had become wealthy and in 1896 he gave the University of Sydney £59,000 to found a school of engineering. It is now the Peter Nicol Russell School of Engineering. Russell's portrait hangs in the Great Hall of the University of Sydney. He was knighted in 1904 and died in London in 1905.

His nephew was John Peter Russell (1858-1931), the Australian Impressionist painter who worked with Van Gogh, Monet, and Rodin. It was his inheritance from the family foundry that enabled John Russell to travel to Europe and pursue his career in art.

The Balmain land had been used by Russell as a depot. Balmain Council acquired the land and proclaimed it as a place of public enjoyment on 8 July 1921. It was named for a former mayor of the district, Reginald Thornton.

Stroll up steep Darling Street

In the early days, Superintendent of Convicts Captain McLean settled at Peacock Point and his cows created a track by wandering back and forth to pastures in Leichhardt. Before he subdivided his land, John Gilchrist made a road which ran the length of the peninsula through his property. Sir Ralph Darling was the governor of New South Wales at the time and Gilchrist named his main street Darling Street.

As early as 1884 it was proposed that steam trams run to Balmain West (now Rozelle). It was 1886 before Balmain Council voted to have a tramline run from the Darling Street wharf to Balmain West, but nothing came of the scheme. A single line tramway finally opened in May 1892 and was an extension of the Forest Lodge line. The steam trams ran from Bridge Road, Glebe, to the corner of Merton and Darling streets, Balmain. The steep grades of Balmain caused considerable problems but in October 1892 the steam trams were extended to Gladstone Park. The trams picked up their coke supplies near the Lillie Bridge racecourse and at the Mt Vernon Street junction. There were also water cranes at Bridge Road, Forest Lodge, and the Balmain terminus. Passengers enjoyed a service now unknown — they could post their letters on the public transport. Each rear car carried a red postal box for the convenience of passengers.

Sydney's tramways were electrified in 1899, and Balmain's steam trams began to be replaced in 1902. The slope of Darling Street from the wharf to Nicholson Street was still too steep for the electric tram cars so a dummy was used. The top dummy ran on the tram lines and the heavy underground dummy also travelled on lines. Both dummies were connected with a wire rope and the tram was steadied downhill and assisted uphill. In 1954 buses replaced the trams that ran from Birchgrove and from the Darling Street wharf to Canterbury. The next year the service from the Darling Street wharf to the city ceased and in 1958 the tram service from Sydney to Rowntree Street also ended.

At the junction of Weston Street stand two old stone buildings. The eastern one is the old Dolphin Hotel (1844) which later became the Shipwright's Arms (1846). The hotel retained its licence until 1966. It was built by a shipwright, John Bell, who purchased the land from John Gilchrist. It was at one time owned by Fenwicks, the tug proprietors. It is said that its cellar was used by young men returning to Balmain on the late ferry on Saturday night to sleep off a good time had in Sydney.

Opposite is Waterman's Cottage. This two-storeyed building dates from 1841 and was built by a Cornish stonemason named John Cavill, one of three part-owners who bought the site from the Gilchrist Trust. There are still shutters at the windows, but the corner balcony is missing. In 1845 an Inspector of Police, William John Wright, bought the house. It acquired the name Waterman's Cottage when Henry McKenzie, a local waterman, lived there between 1880 and 1907. McKenzie operated an emergency after-hours service for ferry travellers, rowing them between Balmain and Miller's Point.

A little up Darling Street on the right is Plym Terrace bearing the date 1884. The terrace of six houses was erected by a builder, John Dobbie, and the terrace has a basement opening to a sunken area. The brick building has a stucco finished (a method of covering exterior walls to resemble stone). Nos 26-28 are 1840s stone twin cottages — note the Scottish dormer windows.

A row of Victorian terraces on the left has a garden of frangipani trees and tall pink hollyhocks. Balmain offers delightful little scenes, an old blue plumbago bush overhangs worn and weathered steps.

Nos 44-46 have rare half and quarter panels of cast iron in the same pattern. Stop and admire no. 50, Cahermore, a lovely stone residence. A window over the door carries its name in stained glass. Built as a hotel in 1846 by Sydney publican Charles Bullivant, it was known as the Marquis of Waterford and the Marquis Arms. The ceilings were painted by artist George Turner when he resided here in the 1920s. Turner lay on his back to cover the surfaces with Australian flora and fauna. The paintings have now vanished.

Proceed to Nicholson Street

On the corner of Nicholson Street was James Beattie's first butcher shop. He built his shop and house in 1851 and is remembered in the naming of Beattie Street. He later moved to the corner of Darling Street and Queen's Place and carried on his business in Balmain for many years.

Turn right into Nicholson Street then right into Lookes Avenue

Joseph Looke (*see* Profile) was a bounty immigrant (a system in which money was granted as a bounty by the governor of a colony to a settler who assisted in the immigration of a skilled person to the colony to Australia in 1832. He brought his wife, Hannah, and their two sons and a daughter to Sydney town where he established himself as a boatbuilder at Darling Harbour. The deep water inlets of Balmain attracted Looke and he bought waterfront land in 1838.

By 1844 he owned cottages, a boat yard, wharf and timberyard. His immigration to Australia had been a wise move. He also owned land at Port Stephens which provided trees for his timberyard and he extended his interests to the coalfields and built a coal yard at Balmain. The coal was needed for bunkering steam ships in Sydney Harbour. Looke built several houses on the rocky ledges of the thoroughfare which finally became Lookes Avenue. No. 13 is one of his surviving buildings. It was constructed *c.*1840 and was known as Radcliffe in the 1880s. Of one-and-a-half storeys, the dormer windows ended in an unusual parapet. All the detailed stone work has vanished with the passage of years.

Looke lived to build Victorian terraces and between 1866 and 1868 he constructed at 1-5 Lookes Avenue, Alfred Terrace. It seems likely Looke intended to add more houses to this terrace but was prevented from doing so by his death. Sixty-five-year-old Joseph Looke was discovered floating in the harbour close to his wharf on 30 May 1868. Unfortunately many of his properties have disappeared. Cliffdale House, which was opposite Radcliffe, was demolished in the 1970s and Durham Cottage, built *c.*1844 at no. 6 vanished *c.*1965. It had been a Ladies' School in the 1870s. Gone too are the gardens of English flowers, Sweet William, larkspur, verbenas and meadow sweet, mentioned in an earlier century.

Wander down to Brett Avenue on the right

At no. 5 Brett Avenue is Lurley, originally Erith Villa. Built by Sydney merchant John Vinson Barnard between 1856 and 1860, it was purchased by Edward Row, a chemist and druggist, partner in J. & E. Row of Pitt Street, Sydney. Edward Row died at the residence in 1900. His brother John Row bought land and a stone cottage, Alfred Cottage, in Gallimore Avenue. He also built a fine weatherboard home, Devonia House. It had stables and a coach house, but Row lost the house when his business was in difficulty in 1875. The house was demolished in about 1938.

Brett Street is named for an early sailmaker, and 1840s directories disclose local residents' occupations as shipbuilder, boatbuilder, plasterer, sailmaker, and gentleman.

Return to Nicholson Street and walk to the public school

Nicholson Street Public School was designed by William Edmund Kemp (*see* Profile) and celebrated its centenary in 1983. Kemp was a pupil of architect Edmund Blacket and later worked with Colonial Architect James Barnet. His school at Balmain has a free Classicism and is not cluttered with fussy Victorian ornamentation.

Both the Church of England and the Presbyterian Church operated small schools in Balmain in the 1840s. An early Balmain teacher was John Balmain, a second cousin of Surgeon William Balmain of the First Fleet (*see* Profile). For many years he was engaged in the legal battles concerning the Balmain estate.

The old Presbyterian school became a government-supported non-denominational school in the 1850s but it was very small. A new school was built on the Pigeon Ground, now Gladstone Park (*see* p. 39), in 1862 when there were 250 pupils. As Balmain thrived, this school became too small for its pupils and the Adolphus Street School opened in St Mary's Church hall in 1883. In 1876 it was proposed yet another new school for infants and primary be built in Nicholson Street. By the time the school was built and opened in 1883 the Public Instruction Act of 1880 had been passed which provided 'free, secular and compulsory' education in New South Wales. One of the school's most famous pupils is former New South Wales Premier Neville Wran.

The Wrans lived in Darling Street and Neville Wran's great-grandfather was a stonemason. He is said to have worked on the Lands Department, Bridge Street, Sydney and at the General Post Office. Neville Wran's remark, 'Balmain boys don't cry', was made at the time of

Nicholson Street Public School. Designed by William Kemp, it opened in 1883

the Royal Commission set up to inquire as to whether Premier Wran had brought pressure on Chief Stipendiary Magistrate Murray Farquhar to influence the course of justice in a case concerning the Rugby League chief, Kevin Humphreys, another old Balmain boy. Premier Wran said at a Labor Party conference that some of the mud would probably stick but 'Balmain boys don't cry. We're too vulgar and too common for that and probably vote Labor anyway.'

An earlier famous Balmain quote came from a senior New South Wales policeman: 'There are only two kinds of people — those born in Balmain and those who wish they were.'

Return to Darling Street and cross to Johnston Street

On the corner of Johnston Street is the former Unity Hall Hotel, a fine two-storeyed sandstone building. The licence was transferred to the New Unity Hall Hotel on the corner of Beattie and Darling streets in 1874.

Continue down Johnston Street

The view from the junction of Johnston and Paul streets inspired artist George Lawrence, a contemporary of William Dobell, to paint his work The Blue Gate. It hangs in the executive office of the AMP Building in Sydney. Note terraces and a little cottage (*on left*) with turned wooden veranda posts and fanlights in the roof. Opposite is no. 12, which has a carved valance board and rounded iron attics. The house was originally single-storeyed. Built by the Commander of New South Wales Naval Forces, Captain Francis Hixson, in 1865, it was enlarged by Captain John Lyons and named Branksea. It was renamed Onkaparinga by William Robert Snow who came from South Australia. He ran a boarding house here in the 1930s.

The tiny streets are quiet and deserted and much of the colour has gone. In the old days one might have glimpsed the Chinese vegetable man carrying his wares in two large cane baskets hanging from a long pole resting on his shoulder or across his back.

The Johnston's Bay end of the streets was robbed of much of its stone by illegal stone-cutters last century, and in about 1861 the the Council had to fill the holes. There used to be '100 steps' built of sandstone but they vanished when Hosking Street was re-aligned.

Turn left into Pearson Street

The street is named for Captain Pearson and no. 11 was the captain's house. Built *c.*1844, the house was named Eastcliffe by a Major Jaques. Note the iron posts and panels and the carved valance boards to the attics.

Follow St Mary's Street

On the corner of Pearson and St Mary's streets is a block of modern apartments. This was the site of the Church of England rectory. There was no dray access to Peacock Point and St Mary's Street was hewn out of the sandstone when the rectory was built in 1863. The old church rectory was demolished when the apartments were built. There are small cottages of stone and sandstock brick with tiny attic windows in St Mary's Street.

Turn left into William Street

William Street leads to Illoura Reserve and Peacock Point. The point was named for John Thomas Peacock. When Semi-Circular Quay was built in 1846 Peacock collected the wharfage and tonnage dues. Don't miss the sandstock brick house on the corner dating from the 1840s. Note the valance board and tiny cast-iron widow's walk. These 'walks', also called 'captains' walks', gave fine views of the harbour. A captain might keep an eye on harbour shipping or his wife look anxiously for his return from sea, sometimes in vain, hence the name.

Illoura Reserve gives splendid views of the Sydney Harbour Bridge and the sheer rock face of The Rocks area. A plaque states:

THE SITE OF THIS RESERVE WAS USED AS A TIMBER DEPOT FOR MANY YEARS BY THE MARITIME SERVICES BOARD OF NEW SOUTH WALES. IN 1970 THE BOARD CONVERTED THE AREA TO FORM THIS RESERVE AS A CONTRIBUTION TO THE CAPTAIN COOK BI-CENTENARY CELEBRATIONS AND NAMED IT 'ILLOURA RESERVE'. THE FLORA AND DESIGN ARE AUSTRALIAN IN CHARACTER.
THE WORD 'ILLOURA' IS ABORIGINAL FOR PLEASANT PLACE.
OCTOBER 1970.

The park was completed in 1982 and designed by Bruce McKenzie, landscape architect.

The first blocks of land offered for sale in Balmain were on Peacock Point and shipbuilding firms snapped them up. The deep bays around Balmain were ideal for their purposes. Stroll to the lookouts and take in the fine city views. The reserve is well planned with bush settings and a row of she-oaks (casuarina) by the water's edge. Casuarinas are named for the cassowary, as the finer filamentous branches were thought to resemble the bird's quills. The early settlers thought the wood resembled English oak hence the name 'she-oak', also called 'he-oak' or beefwood. The timber was used for bullock yokes, wheel spokes and axe handles. Even today it is used for roof shingles. The shingle roofs of the Mint and Hyde Park Barracks Museums in Sydney are of casuarina.

One can gaze across a tangle of flowering honeysuckle to Darling Harbour or continue around the point to look at the skyline of Pyrmont and White Bay across Johnston's Bay. Pyrmont was named for the German watering spa and White Bay for John White, Surgeon-General of the First Fleet and of the first settlement. White returned to England in 1794 and was succeeded by Dr Balmain.

Square-rigged sailing ships once anchored in these waters and the foreshores were lined with shipyards, boat yards, pharmaceutical works, and light industries. Balmain residents have always taken pride in their peninsula. The suburb had skilled workmen in waterfront trades such as shipbuilding, stevedoring and engineering. For over 100 years these trades were the backbone of Balmain, and the suburb boasted Australia's largest shipyards at Mort's Dock and Cockatoo Island (*see* pp. 27). As well as the maritime industries, there was employment in the electric powerhouse, the coalmine (*see* p. 49), locomotive building, and the chemical industries.

Re-trace your steps and go into Edward Street and Little Edward Street

7-17 Harbour View Terrace was built around the 1860s by a developer, Francis Smith. It is of plastered stone and there are dormer windows and attic bedrooms. No. 23 Edward Street won the RAIA Merit Award in 1978. Built in 1974 by architect Glen Murcutt, the roof has mobile screens that allow controlled light to filter inside. Desmond Villa stood on this site until demolished in the 1960s. It was a brick house with a wide veranda and steps to the shoreline. Built by Francis Smith Jnr in 1871, Desmond Villa was bought and named in 1872 by a city draper and milliner, Nicholson Hopson.

Little Edward Street is a tiny lane and has a row of two-storeyed picked-stone houses (nos 2-8) dating from 1844. Four steep stone steps lead to the front doors. The houses were built by a butcher, William Dennis, who died in an accident at Tunkills Mines in South Australia in 1849. Modern houses stand side by side with old residences. The early stone cottage at no. 10 was built in the 1840s by a Balmain slater named James Suddy.

The little streets give tantalising glimpses of the water. Balmain boys grew up fishing, prawning, catching crabs, scrambling over harbourside rocks, and sailing boats on the harbour.

Turn into William Street

A wooden terrace on the corner is covered with wisteria and is cool and shady. Ada View, another terrace, is pristine white with royal blue cast-iron. Look for the etched-glass panels in the doors of the stone cottages

A stone terrace behind Peacock Point

The maritime flavour of old Balmain still exists

and note the attics in the roofs of nos 25-31. These are Charles' Villas and Elizabeth's Villas, built between 1850 and 1858 by a Sydney publican, John Sims. The cast-iron verandas were added in the 1880s. Charles Rembold, a Canterbury market-gardener, bought them in 1874 and is said to have named them for family members. The original shingle roof is beneath the corrugated iron.

William Street has connections with the Gardner family who ran a shipyard on Johnston's Bay. Henry and Edward Gardner married granddaughters of Thomas Rose, who in 1793 was head of the first free family to arrive in the colony. Henry Gardner built a stone cottage at 18 William Street in about 1841. The Gardners were boatbuilders from 1860 to 1912. Henry was a shipwright and Edward a brass-moulder. The brothers shared the house when it was built, but it later became the home of Henry and his wife Elizabeth Rose who lived there for many years. The house was remodelled in 1962. Henry's brother, Edward, married Louisa Rose and built a neat stone residence at 8 William Street (c.1857), but this was demolished in the mid-1960s. In this period, members of the Rose family left the Hawkesbury and came to Balmain. Joshua Rose built 22 William Street between 1852 and 1860, and Thomas Rose, a shipwright, built 10 William Street in 1857.

Eventually the Roses and Gardners owned all the waterfront in this area of Balmain for their shipbuilding activities.

Turn right into Johnston Street

Johnston Street is a mixture of old and new; boats are pulled up on the waterfront and the eye is continually turned to the harbour.

The peninsula was once covered with tea-tree scrub and gums growing in the rocky soil. The harbour foreshores were thick with oysters and mussels. The Aborigines feasted on these and left behind middens (a refuse mound of shells), evidence of their feasts. Before early settlement kangaroo hunts were conducted on the peninsula but with settlement Balmain grew into a tiny hamlet. In 1851 the population was 1,397; by 1856 it had reached 2,224. It was still bushy: the bushranger Captain Thunderbolt escaped from prison on Cockatoo Island and disappeared in the 'Balmain bush'. Thunderbolt (Frederick Ward) was born at Windsor, New South Wales. After his escape, he carried on bushranging between Newcastle and the Queensland border from 1864 until 1870. He was shot at Uralla, New South Wales, on 25 May 1870.

Take left turn into Hosking Street

Here the modern houses have adopted the features of the old in a happy mixture of the centuries.

Climb the worn stone steps to Union Street

Stop to look at the house of Cornish stonemason John Cavill (*c.*1853) at no. 17 Union Street. Philip Gidley King, First Fleeter and third governor of New South Wales, was the first Cornishman of note to arrive in Australia. Convict James Ruse, who sowed the first wheat in the colony, was another native of Cornwall. John Cavill built many of the stone buildings in Balmain from the late 1840s and lived at 17 Union Street until 1867. He was, no doubt, hard-working and law-abiding like the Cornish miners who flocked to the South Australian copper mines in the 1860s and built towns of sturdy stone houses. Don't miss Elizabeth Terrace (1885) before turning left into Darling Street.

Balmain has traditionally been known as a working class suburb. But the residents were not solely working class — there were sea captains, merchants, and industrialists. Premier Sir Henry Parkes lived in East Balmain, architect Edmund Blacket was another resident, as was Prime Minister 'Billy' Hughes, and Dr H. V. Evatt, QC, politician, historian and president of the United Nations Assembly. Former Governor-General Sir John Kerr was born in Balmain, the son of a boilermaker.

By 1860 Balmain was a municipality and boasted a population of 3,482 which climbed to 6,272 in 1871; 15,663 in 1881; and 23,475 by 1891.

Note Glentworth (*c.*1842) at no. 86 Darling Street. The house was enlarged for Captain John Broomfield in 1888. There is a colonial fanlight over the main door and a cast-iron railing with a fine fern decoration. English houses were surrounded by cast-iron railings before our first settlement in 1788. Foundries were established in Australia and the Hobart Town Foundry and Smithy was supplying decorative panels and railings by 1835 (*see* p. 11). By the 1860s most of the iron used in Sydney came from local manufacturers. Australians now value our decorative iron-work and older suburbs contain many fine examples of the art.

The fern pattern on no. 86 is a graceful design and similar patterns were manufactured in both New South Wales and Victoria around 1885 (NSW) and 1884 (Victoria).

The old stone building on the corner of Darling and Datchett streets was once bootmaker John Cornell's shop and dates from 1860. In hard times and during the Great Depression many of the local children were barefooted. Louis Stone's novel *Jonah* (Angus & Robertson) concerns the fortunes of a street tough who becomes the proprietor of a great shoe store. Jonah deserts his employer, Paasch, to set up his own business and Paasch has to dress his own window:

> But the pride of the collection was a monstrous abortion of a boot, made for a clubfoot, with a sole and heel six inches deep, that had cost Paasch weeks of endless contrivance, and had only one fault — it was as heavy as lead and

Worn stone steps lead up to Union Street

unwearable. But Paasch clung to it with the affection of a mother for her deformed offspring, and gave it the pride of place in the window. And daily the urchins flattened their noses against the panes, fascinated by this monster of a boot, to see it again in dreams on the feet of horrid giants. This melancholy collection was flanked by odd bottles of polish and blacking, and cards of bootlaces of such unusual strength that elephants were shown vainly trying to break them.

Cross Darling Street to view St Mary's Church

The first Anglican church in Balmain was a slab and bark building near Waterview Bay (now Mort Bay). St Mary the Virgin is showing signs of age as the sandstone is badly eroded and worn. Architect Edmund Blacket suggested the design in 1843, but building did not commence until 1845. It is one of his earliest buildings. Thomas Sutcliffe Mort of Mort's Dock fame (*see* p. 27) was a great churchman and also a friend of Blacket's. Mort was living in Ewenton Street when St Mary's was extended in 1859 and he gave the church the chancel window which depicts Solomon, David and Christ. The extension was carried out by former Colonial Architect William Weaver and by William Kemp, architect of the Nicholson Street School.

St Mary's parish originally included Balmain North, Balmain West (later Rozelle), and Ashfield. Blacket's distinctive 13th-century decorated Gothic architecture is still discernible (the minister, Mr Wilkinson, wanted a Norman church) and the old church stands aloof from the traffic passing up and down Darling Street. There is a row of narrow terraces at nos 103-109. Land was expensive so builders endeavoured to fit as many homes as possible on a block. Before leaving Darling Street note Pinetree House no. 122 on the opposite side of the road. It dates from the 1850s and has a slate roof.

Turn right into The Avenue

The Avenue leads to Mort Bay. The old days of Balmain are only a memory. The *Balmain Almanac* of 1878 declared 'Larrikinism has to a great extent subsided.' The larrikins had grown to manhood but the Almanac noted 'they still prefer the gutters to sit in'. Nevertheless in the 1920s and 1930s there were still plenty of boys around to seize the opportunity to plunder fruit-trees growing in suburban gardens. An irate householder often pursued the thief down the streets of Balmain.

The first New South Wales Labor Electoral League was established in Balmain on 6 April 1891, the forerunner of the Australian Labor Party, and the suburb has always had a strong Labor working-class community. In the 1930s people moved to the district seeking cheap rents. The early 1930s was the era of the Great Depression and Balmain suffered the hardships of those grim years. The value of our exports fell by half, farms were abandoned, shops closed, families were turned onto the streets and at one stage thirty per cent of the workforce was unemployed. The Depression left an indelible mark on all who lived through those years. Men wandered the streets offering shoe laces, soap, needles, rabbits, bunches of gum tips, anything for sale. Shanty towns sprang up around Sydney — shacks of bagging and waste timber to house evicted families. Lady Game, wife of Governor Sir Philip Game, visited one such settlement at La Perouse. On her return to Government House she stated she had seen one little house in which she would not mind living herself! In 1931 the basic wage was cut by ten per cent, and in 1933 one-quarter of the workforce was unemployed, but the Depression had begun to ease.

It was a time of highs and lows for Australia. In 1931 artist Tom Roberts and singer Dame Nellie Melba died. The racehorse Phar Lap died in the United States of America in the following year. 1933 saw Australia retain the Ashes following the controversial 'body-line' cricket tour. The people of Balmain, like all Australians, rejoiced in the triumphs of Don Bradman. When the South African team visited Australia in 1931-2 Bradman made a century in every test he batted in — his best score 299 not out. Sydneysiders boasted of 'Our Harbour — Our Bridge — Our Bradman'. Balmain survived the Depression and its men went off to the war in 1939.

At war's end the old Balmain was changing. By 1947 Balmain came under Leichhardt Municipal Council and there were schemes of demolition and re-building in the suburb. It was decided to banish the narrow lanes and build wide roads. In the 1960s public housing in the form of flats was built, and red-brick blocks replaced old homes. But the suburb is close to the city and some realised its potential and began to restore old cottages and houses. Balmain became 'trendy'. The Avenue has terraces and small cottages to admire and leafy green trees provide shade on a hot summer's day.

Continue down to the harbour

Thomas Sutcliffe Mort arrived in Australia in 1838 and by 1843 had his own wool-selling agency. He had varied interests but Mort's Dock is his most famous memorial. The idea of a dry dock came from Captain Thomas Stephenson Rowntree (*see* Profile). The captain, from Durham, England, went to sea in 1838, the same year Mort arrived in Australia. Rowntree rose from ship's carpenter (he was a shipwright by trade) to Master in four years. He built a ship, the *Lizzie Webber*, and brought passengers to the Australian goldfields. The vessel was moored in Waterview Bay (now Mort Bay) and Captain Rowntree settled in Balmain in 1851. In 1853 he proposed a patent slip on his property at Waterview Bay

Mort's Dock in 1874

but, after approaching Mort to auction the *Lizzie Webber,* the pair decided to construct a dry dock. The foundation stone was laid in 1854 and Mort, Rowntree, and J. S. Mitchell formed Rowntree & Co. Rowntree later left the enterprise but went on to become a foundation alderman of Balmain Municipality and was twice mayor of the area. The dock was originally Waterview Dock and the first ship to enter the dock was a steamer *Hunter* in February 1855. The dock was building ships the same year and in 1870 was constructing locomotives. In 1872 the Mort's Dock & Engineering Co. was incorporated.

Mort was able to overcome competition from the Government Dock by persuading Governor Sir Charles FitzRoy to restrict it to working on naval vessels. Mort brought workers into the area to build the dock and promised them a free block of land upon its completion. It is not clear if these were to be individual blocks or one common block or whether the promise was fulfilled. Mort owned nearly all the Waterview area. He died on 9 May 1878, and in April Balmain Council named the bay Mort Bay in his honour. The dry dock handled English and North American mail steamers and as the size of shipping increased heavier machinery was installed at the dock. By 1883 some residents complained of pollution from the dock and there was an outcry when some of the streets (Yeend and Mort streets which led to the ferry) were threatened with closure. Industrial trouble had developed at the dock in the 1870s when workers wanted an eight-hour day instead of a six-day week of twelve hours. A strike in 1873 ended when the workers won their eight-hour day.

Balmain played an important role in union matters and the Mort's Dock workers grew in militancy. Dock conditions were often hazardous and dangerous. One worker fell down a well in 1883 and was killed. By the 1890s there was a demand for only union labour at the dock.

In 1901 Mort's Dock & Engineering Co. opened a graving dock at Woolwich, a mile (1.6 km.) from Balmain, and there were two floating docks at Jubilee Bay, Balmain. Mort's Dock was one of the greatest industrial centres in Australia. Many of the earlier Manly and inner harbour ferries were built there. The Great Depression took its toll at the dock and in 1932 half of the workers had lost their jobs. Employment did not improve in the area until World War II. At war's end financial problems plagued the company in the 1950s. In January 1959 Mort's Dock & Engineering Co. went into liquidation and the Balmain and Woolwich plants closed. In 1965 the old dock was demolished and the dry dock filled for new wharves. A new era commenced when the first container ship berthed there in 1969, although there was a public outcry about the heavy container trucks using the narrow Balmain streets.

A $40 million housing development on the old Mort's Dock site is built over the tunnel to the old Balmain coalmine (*see* p. 49). The old tunnel is some 900 metres under the surface — the coalmine was one of the deepest in the world. The Housing Department denies there is any evidence of subsidence or any danger from methane gas emanating from the old mine.

The area is full of Sydney harbour ferries as the ferry maintenance depot is situated in the bay and there are scenes to entice the marine artist. One cannot fail to notice the huge Colgate-Palmolive factory on the foreshore. In 1895 Lever Brothers, makers of Sunlight soap, opened a mill at Balmain to extract oil from copra and in 1900 began to manufacture soap in a factory next to the mill. The Colgate-Palmolive factory has provided employment for many Balmain residents but is due for demolition. Across the bay great earth-moving machines cleared disused areas and government housing covers the site.

Follow shoreline west

This area has been developed with new housing. Some are modern in style while others have adopted the best features of the old architecture. Great use is being made of timber lattice in the new Balmain housing. In the old days the use of timber for housing meant jobs and the natural wood is an excellent material for the modern residences. Wander up Hart Street and study the new architecture. The road leads back to Balmain's main thoroughfare, Darling Street, and the red-brick church of St Augustine dominates the skyline (*see* p. 37). Wander up to the Watch House, commenced in 1854 by architect Edmund Blacket as a single-storeyed lock-up. It was restored by the Balmain Association (open Saturdays 1-9 pm) (*see* p. 31). If you are keen to see more of Balmain, Walk 2 offers a different view.

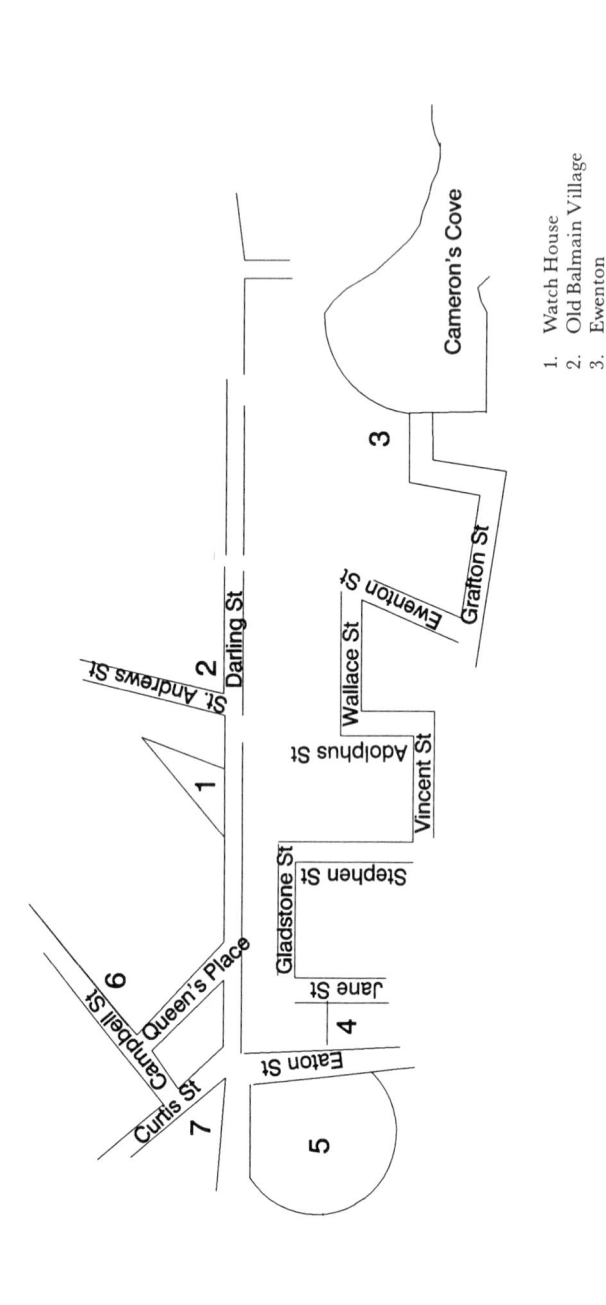

Walk No. 2

Balmain

From the Watch House to Cameron's Cove, St Augustine, Gladstone Park, Queen's Place and St Andrew's Congregational Church

Commence walk at the Watch House, Darling Street

The Watch House. The Balmain Association lovingly restored Edmund Blacket's old lock-up in 1970 and the building is now their headquarters. It is uncharacteristic of Blacket as the style is Georgian. It served as a police station and housed six police officers and also had four cells and two exercise yards for prisoners plus a charge-room and kitchen. Blacket also built a twin building at Victoria Cross, North Sydney, which has now vanished. A privy at the rear of the Watch House was built in 1855 by William Kemp (*see* Profile), architect of Balmain's Nicholson Street School.

The top storey of Balmain's Watch House was added in 1881 and provided living quarters. A new police station and court house were built in 1887 and the Watch House's role was then purely residential. In the 1950s it fell into disrepair. The Balmain Association fought against its demolition and restored it for the Captain Cook Bi-Centenary celebrations.

A local publication, *Bally-Who*, in February 1987 suggested the past still lingered at the Watch House. It told of ghostly footsteps echoing down the wooden veranda although no one could be seen. As day drew to a close a cold area was experienced near a male cell door. A female figure appeared on the stairs and vanished into a first-floor bedroom. *Bally-Who* could find no evidence of horrific events at the Watch House in times

The Watch House. The old lock-up is now home to the Balmain Association

past, but record-keeping was scant between 1850 and 1890.

In a grassed area close to the Watch House are the foundations of the old Presbyterian Church. The land was bought in 1841 by John Dunmore Lang for a Presbyterian church. In the 1850s the Reverend Thomas Gordon endeavoured to join the Presbyterians and the Congregationalists in one body but the two congregations separated in 1857. The small Presbyterian Church ceased to be used and a new weatherboard church was built in what is now Colgate Avenue. There is a fine stone Presbyterian Church in Campbell Street designed by James McDonald, one-time mayor of Balmain.

Walk down Darling Street

Travel east down Darling Street and note the stone pediment on no. 177. You will pass the corner shops and cottages of the old Balmain village built in the period 1845-6. Note the buildings in St Andrew Street — the corner shop is easily identifiable. These small shops were like community centres in earlier days. Residents met each other here and exchanged news and local happenings. The goods were plain and simple, tea, sugar, flour, treacle, butter, a paper twist of boiled lollies for a child were carried home in a strong cane basket.

In Cooper Street just around the corner from Darling Street there is a tiny stone cottage. A glimpse of harbour can be seen at the end of the street. On the right is Balmain Bowling Club, established 1880, the oldest bowling club still playing on its first site in Sydney.

Continue down Darling Street to the right hand turn to Cameron's Cove

New town houses have attracted more residents to old Balmain in the development at Cameron's Cove by Lend Lease Corporation Ltd. Many of the houses have stunning views across Johnston's Bay to the city. The bay was named for George Johnston, Captain of Marines in the First Fleet (*see* Annandale walk). The tall chimney stacks of the Pyrmont powerhouse were a landmark of the area for many years. Close by the new development is a home of a different generation. This is Ewenton — for many years it was a ruin and was gutted by fire but has now been fully restored externally. The house was commenced about 1854 by builder Robert Blake, and named Blake Vale. In 1856 it was bought by Major Ewen Wallace Cameron, a partner of Thomas Sutcliffe Mort (*see* p. 27), vice-president of Sydney Hospital and a distinguished citizen of Balmain. Cameron engaged James McDonald to add an extra storey and entrance porch to the original home. The architect added an extra wing in 1872 and a fine bay window gave views to Sydney. The house is built of polished Pyrmont sandstone and the two periods of construction are clearly visible. Major Cameron lived at Ewenton, as he renamed his residence, until his death in 1872. From 1882 to 1891 surveyor Charles C. Cameron resided at Ewenton. He also was an associate of T. S. Mort. The house then suffered many vicissitudes and was sold in 1892 and used by various institutions. From 1893 to 1895 it became a boarding house run by a Mrs Stainger, in 1897 it was the Bethany Deaconess Institution; and in 1916 it became the home of Henry B. Swan, Mayor of Balmain in 1893, 1894, and 1915. Barrister and mayor L. B. Swan lived at Ewenton between 1929 and 1931.

In the 1950s the house was threatened with demolition but despite a fire in 1980 Ewenton survived. It was now been externally restored by the Lend Lease Corporation and its windows face the waters of Cameron's Cove.

From Ewenton proceed to Grafton Street

Grafton Street leads up behind the White Bay container terminal and there are views across to Glebe Island. White Bay was the destination for the seaborne coal trade. 'The Sixty Milers', so-called because it was sixty miles (97 km.) from Sydney to the coal port on the Hunter River at Newcastle, carried 1.5 million tons of coal per year. Residents of the northern beaches were familiar with the 'Sixty Milers' as they ploughed back and forth between their ports. This was a busy part of the harbour in the 1880s with both Johnston's Bay and White Bay crowded with shipping. A sugar refinery was established on the Pyrmont Peninsula at Johnston's Bay in 1878. Transporter cranes handled the raw sugar in bulk; it was not

JOHN BOOTH & CO. LTD.,
TIMBER MERCHANTS,

Sawn Pine, Kauri, Cedar, Hardwood, Door Sashes, and all kinds of Joinery. Importers of Building Material, Glass, Iron, &c. Country Orders promptly executed. Estimates given.

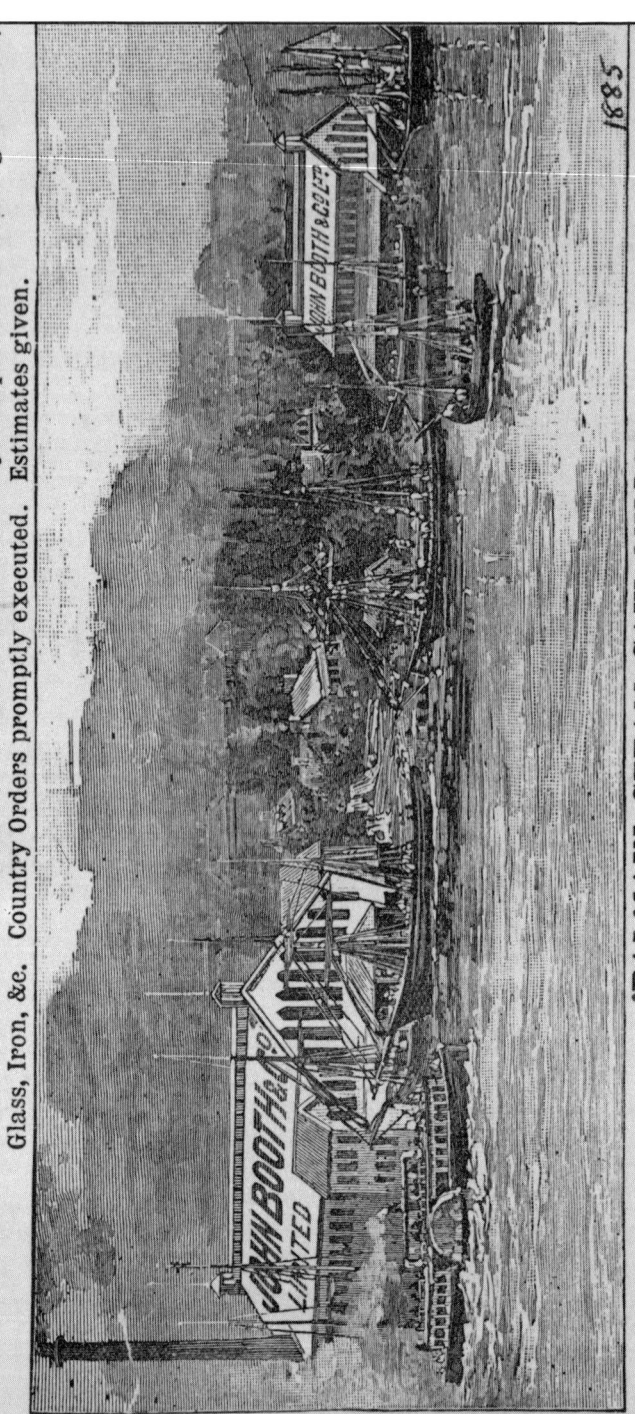

"BALMAIN STEAM SAW MILLS."

BRANCH ESTABLISHMENTS:- MARKET STREET, MARKET WHARF, SYDNEY, AND AT SUMMER HILL. STEAM SAW MILLS: MANNING RIVER

bagged. The works included a sugar refinery, a distillery, an engineering workshop, and a factory making wallboard from pressed sugar cane. The bulk sugar was shipped down from Queensland and northern New South Wales. The bulk of the raw sugar is still transported by sea but motorised transports are also used. There is still a distillery at Pyrmont plus a central laboratory and an engineering training and apprenticeship branch. Building materials are manufactured now by the Gypsum Products Group at Wetherill Park.

The streets overlook the bays and there are old weatherboard houses. Balmain has the highest percentage of 19th-century wooden houses in Sydney. Stop and look at Hampton Villa at no. 12B Grafton Street. It was built in 1855 by the Honorable Edward Hunt and leased by State Premier Sir Henry Parkes in the 1880s. Hampton Villa has undergone extensive restoration and additions.

Turn right from Grafton Street into Ewenton Street

Ewenton Street has two surviving houses built by Robert Blake, the first local builder and creator of Ewenton. Blake purchased eight acres thirty perches (3 ha.) of land in 1837. He built Shannon Grove, no. 10 Ewenton Street, in 1848 for £500. The house has stringy bark floors and cedar joinery. Attic bedrooms were added in the 1890s. Blake was one of the first developers on the point and he built another eight houses. Between 1880 and 1887 Shannon Grove was the home of John Cameron, Mayor of Balmain from 1883 to 1884.

Blake's other surviving house in Ewenton Street is Kinwarra, no. 3, completed in 1852.

Turn left into Wallace Street

No. 1 Wallace Street is known locally as the Railway Station. The rough stone two-storeyed house somewhat resembles a country railway station. It stands on land that once formed part of the estate of builder Robert Blake and dates from the late 1870s. Behind is a small house, Moorfield, built in 1838 by John Bibb for Robert Blake. It has been claimed that Bibb had been a pupil of John Verge, the architect of Camden Park House and Elizabeth Bay House. Bibb also built the Union Bank (now demolished) and the Congregational Church in Pitt Street, Sydney.

On the corner of Wallace and Adolphus Street there is now government housing for senior Balmain citizens.

Turn left into Adolphus Street and right into Vincent Street

In Adolphus Street are small stone workmen's cottages. Life was hard for a working class housewife of an earlier generation. There were no labour-saving devices. Clothes were boiled, rinsed, blued, and starched. Ironing was done with a heavy flat iron heated on the stove. The wooden kitchen table was scrubbed with sandsoap to make it white; the kitchen grate was blackened. Money was always scarce but food, clothing, and shoes had to be bought, so if a husband were unemployed or took to drink it was disastrous for his wife and family as there were no pensions.

In the 1830s and 1840s there were several temperance journals in Sydney and in the early 1900s cordial manufacturers pushed raspberry vinegar and raspberry syrup as temperance drinks. Mostly these were consumed by women and children. The New South Wales Government Analyst said 'with a little less sugar, these liquids might be used as red ink'. The mixtures were mostly composed of coal-tar dye, cochineal, salicyclic acid, and saccharin but not a trace of raspberry juice. Another temperance drink, hop beer, contained four per cent alcohol! On the corner of Adolphus and Vincent streets is the former Rob Roy Hotel (1855). It would have echoed to the sound of workmen's voices. A heavily loaded dray brought wooden barrels of beer to be man-handled and rolled into the hotel cellar. In the night the hotel lights enticed passers-by into its warmth and the local policeman passed on his beat to make sure there were no rowdy brawls. In the 1870s Balmain held its policemen in high esteem. The *Balmain Almanac* of 1878 states:

> Our police, 6 in number, are amongst the most exemplary that New South Wales can produce; they are more parental than otherwise — protective — not detective — but they always look to the public interests.

Continue to no. 7 Vincent Street

The Grange was the home of William Adolphus Young (remembered in Adolphus Street), High Sheriff of New South Wales from 1842 to 1849. Young's second wife was Jane Throsby, a daughter of Charles Throsby, the early settler and pioneer. Young sailed to England in 1850 and between 1857 and 1875 was a member of the British House of Commons. The Grange is a simple Georgian cottage dating from 1842 and was restored in 1966-7. At one stage the old house faced a demolition order. There are shutters to the long veranda windows, attics in the roof, and urns of flowering plants.

Leave Vincent Street for Stephen Street on the right

White geraniums poke through iron-railing fences. The corner hotel in Stephen Street is decorated with tiles, a survivor of the forty hotels that once dotted Balmain. The streets were narrow in Victorian Sydney and full of rubbish trampled by passing horses. The death rate was high and twelve per cent of all one-year-old children died of scarletina, fever, and teething problems. The air was said to be 'infected with the vilest odours' and the harbour polluted 'beyond endurance'.

Turn left and climb the hill of Gladstone Street

Gladstone Street is narrow and there are small stone and timber houses with picket fences to glimpse. Back down the hill there are views of Balmain and the harbour. On the wall of one old cottage yellow and orange nasturtiums climb the stones and English flowers grow in a small garden. It is a steep climb to Jane Street.

Turn left in Jane Street

At 14 Jane Street is the Aboriginal Training and Cultural Institute. This building was formerly the Convent of the Immaculate Conception, designed in 1876 by architects Edmund Blacket and J. Horbury Hunt. Blacket arrived in Australia in 1842 and his only training as an architect was his hobby of drawing and measuring English churches. J. Horbury Hunt has been called the architect's architect and he designed the original New South Wales Art Gallery in the Sydney Domain.

Across the road from the convent is St Augustine's Church, a small stone building overshadowed by the newer brick St Augustine's. The land was given to the Catholic community of Balmain by the High Sheriff of New South Wales, William Adolphus Young. It is a magnificent site, high on the hill of Balmain, with sweeping views of the area. The stone church, which now serves as a parish hall, was designed by J. F. Hilly. The foundation stone was laid on 4 September 1848 and the church opened on 5 September 1851. In 1860 it was extended by twelve feet (3.7 metres), and a gallery was added.

Hilly was the architect of the elegant Royal Exchange in Bridge Street, Sydney, which was demolished in the 1960s. The church has a connection with the beginnings of the Catholic faith in Australia for the Reverend John Joseph Therry was parish priest of Balmain. He died at the old presbytery on 25 May 1864. Father Therry, first official priest in Australia, arrived in Sydney on 3 May 1820 and through the efforts of Governor Lachlan Macquarie laid the foundation stone of the first Catholic Church, St Mary's, in Sydney in 1821. When Archbishop Polding arrived in Sydney in 1835 Father Therry became a parish priest and served in Campbelltown (New South Wales), Hobart (Van Diemen's Land), Melbourne (Port Phillip district), Sydney, and finally Balmain in 1856. His tomb is in the crypt of St Mary's Cathedral, Sydney.

The modern St Augustine's is a Balmain landmark and was designed

Looking east down Darling Street from an entrance to Gladstone Park

by architect E. A. Bates in an art nouveau style. Its foundation stone was laid in 1906 and the church dedicated in 1907.

Walk from Jane Street through the church grounds to Eaton Street and Gladstone Park

Gladstone Park was originally the Pigeon Ground as pigeon shooting took place here in the 1850s. The land was privately owned but the Balmain citizens wanted a park. It took many years of agitation before the Pigeon Ground was proclaimed in 1885 but the park was not officially opened until 5 April 1890. It was enclosed by a white picket fence and named Gladstone Park in honour of the British prime minister. But by 1905 it was described as 'a dusty heap of stones' and an 'offensive tip'. Some of the park was taken by the Department of Education for school grounds. The old Pigeon Ground School was founded in 1862 — the lower storey of the Father Michael Rohan School, in the grounds of St Augustine's, is part of this school. It was not on the Pigeon Ground but faced it and thus acquired its name.

The Bowling Club was founded in 1898, but the public was granted access to the grounds. In 1912 there was a public gymnasium but it was demolished in the same year because of a number of accidents.

Also in 1912 it was announced a water reservoir was to be excavated under Gladstone Park. Like Sydney and many of its environs, Balmain faced water restrictions as the suburb grew. In 1884 Balmain residents had been allowed water only on Tuesdays, Wednesdays, Saturdays, and Sundays. The rock under the park was cut to a depth of eighteen feet six inches (5.6 m.).

The reservoir was one hundred and eighty feet long (54.8 m.) and one hundred and five feet wide (32 m.) Excavation was completed in 1915 but in 1916 Balmain was entirely without water during a period of drought. By June 1917 the Gladstone Park Reservoir was ready for use and there was a pumping station in Booth Street. The reservoir holds 2.4 million gallons (10.9 million l.) and is now used only for standby purposes. The Water Board had to restore the park on completion of the work and provided a bandstand which opened on 7 April 1918. The bandstand was demolished in 1951 because of its neglected state.

The pigeons are still in Gladstone Park but are no longer shot. The area is a refuge for Balmain residents with its leafy green trees and flower beds. Note the old cast-iron posts in the park.

From Gladstone Park return to Darling Street

No. 238 Darling Street was originally the Bank of New South Wales in 1875. The bank was established in Macquarie Place, Sydney, in 1817 by

Governor Lachlan Macquarie and although it suffered initial problems it grew to be Australia's largest trading bank. The bank still functions today as Westpac.

No. 236 Darling Street is the Oddfellows Hall dating from 1864. The Workingmen's Institute used the basement as a meeting room in 1865. A new Workingmen's Institute was built between 1893 and 1896. The building, further up Darling Street, has been restored as the Institute Arcade with shops and offices.

The Albion Hotel at 212 Darling Street dates from 1860. Lown's grocery shop at no. 208 and Jung's bakehouse at no. 210 dated from the 1850s. On the corner of Darling and Jane streets is the London Hotel and at 206 Darling Street is an 1850s tobacconist shop.

Cross Darling Street to Curtis Road

At the junction of Darling Street and Curtis Road is an 1868 shop constructed by architect James McDonald for George Verey, draper. James McDonald also designed the Presbyterian Church in Campbell Street and was once mayor of Balmain. The wooden shelves of the shop would have been crowded with bolts of material — ginghams, cottons, calico, silks; and cards of buttons, fasteners, and elastics. Perhaps Verey's also carried gloves and umbrellas, tussore silk parasols lined with green, sky, pink, navy or red for ladies' and maids' sunshades of black moire or black broche. There were gloves of lisle or Ladies' Long Silk Mitts —

> INSTRUCTIONS FOR PUTTING ON GLOVES.
> Do not put them on in a Hurry.
>
> Turn the wrist well back before inserting the hand, then gently work the fingers on and afterwards the thumb also.
>
> Never force the fingers into place by pressing between them with the fingers of the other hand.
>
> Never try to wear a glove which is too small for you.
>
> Carefully follow these instructions and your gloves will seldom split.

From Darling Street walk into Queen's Place

On the corner of Queen's Place and Waterview Street is the Balmain Post Restaurant. This was earlier a corner shop built by grocer Alexander Chape between 1844 and 1857. Chape became the Balmain postmaster in 1857 and he sold money orders from his premises in 1866. After his death in 1870 his widow, Catherine Chape, ran the business and had an agency for the Government Savings Bank at the store in 1871. The corner shop has also been a store and wine bar.

Queen's Place contains an interesting collection of small houses, nos 5-9, survivors of the 1850s.

One building is stone but the other two are timber and a former resident of one cottage was Catherine Chape. It is said a pawnbroker and a carpenter rented the other two buildings.

From Queen's Place continue right to Campbell Street

The solid stone Presbyterian Church is the work of architect and Mayor of Balmain James McDonald. It was built between 1867 and 1868 and follows the Scots tradition of having separately roofed nave and aisles. William and James Burt, local residents, did the stonework of the church. William Burt was responsible for Balmain's Congregational Church. The church has an octagonal tower and the adjoining manse was built in 1905.

From Campbell Street return to Curtis Street

In Curtis Street is the Congregational Church, St Andrew's. The land was a gift of Robert Ford and the stonemason, William Burt, quarried the stone in Balmain. The first Congregational services were held in 1853 in a timber building at the corner of Darling and Cooper streets until St Andrew's was built in 1855 by architects Goold and Field. The church hall is by James McDonald and dates from 1871. There were two Congregational missionaries preaching in New South Wales in 1798 'in different places in the country about the town of Parramatta'. They were then known as 'Independents' a title they inherited from 17th-century Puritans. The founder was Robert Browne (1550-1633), a Church of England clergyman and schoolmaster. He became a preacher in London and wrote tracts on the theme that religion is a matter for the conscience of local congregations and should not be confused with politics. Any body of faithful Christians constituted a church. His followers were persecuted by Elizabeth I, some hanged for treason, others imprisoned. The survivors worshipped in secret and emigrated when times were bad. The Pilgrim Fathers were Independents who fled to America in 1620. Congregationalism is a leading denomination in the United States of America and the great colleges of Yale and Harvard were founded by the movement.

On Saturdays the Balmain market operates in the church grounds. On occasion, brides have had to wend their way to the church through busy stalls and crowds.

End your walk by enjoying refreshments at one of Balmain's cafes.

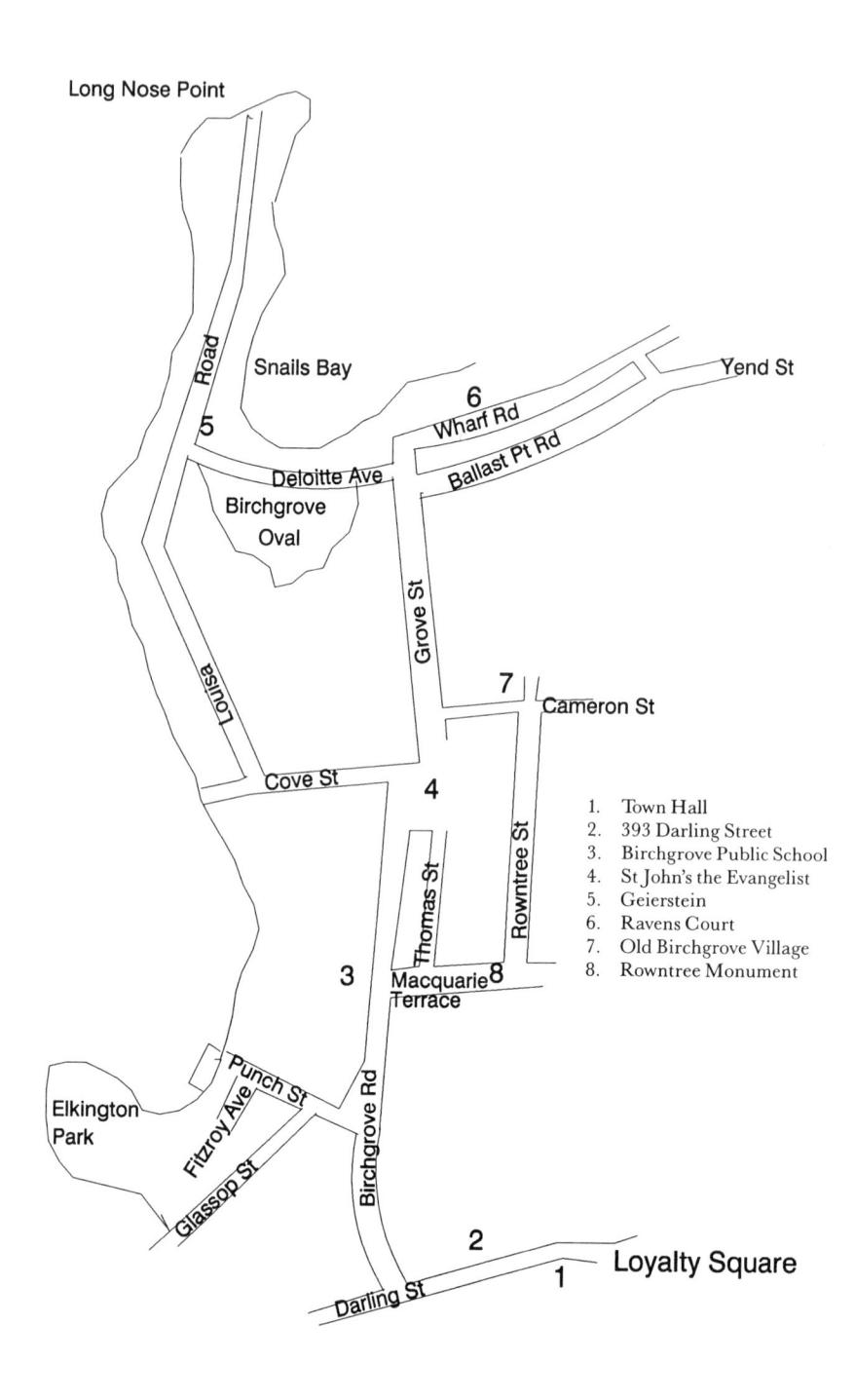

1. Town Hall
2. 393 Darling Street
3. Birchgrove Public School
4. St John's the Evangelist
5. Geierstein
6. Ravens Court
7. Old Birchgrove Village
8. Rowntree Monument

Walk No. 3

Birchgrove

(*Note* — This walk is rather long and it is wise to carry some refreshments to enjoy on Long Nose Point. The walk can be terminated at Long Nose Point and a ferry caught back to the city. Alternatively, after a rest on the point, the walk resumes and eventually leads back to Darling Street.)

The walk to Birchgrove commences at Loyalty Square, the junction of Darling, Beattie, and Palmer streets. This was once Unity Square named for the Unity Hall Hotel. In the centre of the road junction is a memorial commemorating the soldiers of the Great War 1914-1918. The heavy traffic of Balmain roars by the drivers oblivious to the names of thirty-eight Balmain men who gave their lives for their country.

William Morris Hughes was a Balmain resident in the early 1900s — he lived in nearby Beattie Street. Hughes was born in London on 25 September 1864 to Welsh parents. In 1884 he emigrated to Brisbane but moved to Sydney in 1886. Hughes later had a shop in Balmain where he mended locks and umbrellas and sold books dealing with social and political questions. At the time of the 1890 maritime strike Hughes counselled members of the Wharf Labourers' Union, many of whom lived in Balmain, and was invited to become union secretary. He later formed and was president of the Waterside Workers' Federation. William 'Billy' Hughes became prime minister in 1915 but in his early political life is said to have held his meetings in the old Unity Square. (*see* Profile).

Walk west in Darling Street to the Town Hall

On 21 February 1860 Balmain was incorporated as a municipality. The first council meeting was held in a loft of a store owned by Captain Rowntree (*see* p. 57 and Profile) at Waterview Bay. Various sites were used for council meetings until the site of the Town Hall was purchased in 1876. A town hall was built in 1881 by the busy Balmain architect, James McDonald. It still stands at the rear of the 1888 Town Hall. The architect of this substantial Victorian building was Alderman E. H. Buchanan who designed a dome to the town hall but, like Sydney's GPO tower, the dome was removed during World War II as it was feared it would cause injury during air raids. Unlike the GPO clock tower it has not been replaced. By act of parliament in 1948 Balmain Council was amalgamated with Leichhardt Council and the Balmain Town Hall now houses offices and a municipal library.

The complex of post office, police station and court house is the work of Colonial Architect James Barnet who designed many Sydney buildings including the Lands Department in Bridge Street and the Martin Place General Post Office. Architect Edmund Blacket had employed Barnet as his Clerk of Works while building the Great Hall of Sydney University. Barnet was colonial architect from 1865 to 1890 and he designed the Balmain buildings between 1885 and 1887. In the tiny green area between Balmain Court House and Town Hall there is a small glass-roofed plant house where council gardeners cultivate plants for the municipal gardens. The old damaged roof of the post office tower was removed in 1957 because of 'public risk' but restored in the 1970s. This group of buildings are substantial reminders of the economic boom that swept through Sydney in the 1880s. It was then that lifts were first used in Sydney's buildings and all over the city and suburbs architects and artisans were busy creating great stone edifices. Barnet's Balmain buildings have a flavour of India and somewhat resemble his court house at Bathurst, New South Wales.

Opposite the Balmain Court House is the Balmain Fire Station. At the top of the facade is clearly printed

M.F.B.
ERECTED
1894.

Look for the fireman's helmet decoration.

The building cost £1,022 in 1894 and was erected by the Metropolitan Fire Board for the Balmain volunteer fire-fighters. The volunteers manned the station until 1901 when permanent staff were appointed. Soldiers and convicts were the first fire-fighters in the colony and the first full-time services were provided by insurance companies anxious to protect their clients' properties. It is not unusual to still find on an old building an insurance company plate, or fire-mark. Sometimes when the fire service arrived at the burning building they refused to quell the fire if the premises were insured by a rival company.

The volunteer fire services were often given financial help by local government. It was always a problem to get enough water to the fire. The early fire engines were horse-drawn and the first steam fire engines did not appear until the 1860s. In the 1880s telephones were connected to fire stations. The Fire Brigades Board of Sydney was established in 1884 and the volunteers were registered and worked with the permanent brigades. In 1902 in Sydney there were thirteen metropolitan stations with 113 permanent officers and firemen and nineteen volunteer fire brigades, with eleven steam engines, twenty-one manual engines, nineteen horse-carriages and eighty-five horses.

Walk to no. 393 Darling Street

This was the home of Edmund Blacket. Sarah Blacket, the architect's wife, died in 1869 and Blacket wished to leave their home Bidura in Glebe for a plainer house. He had purchased the land in September 1860 for £200 and built his new house about 1870 but by 1875 had moved to Alderly in Booth Street, later part of Balmain Hospital. No. 393 became the Brussels College for Ladies in 1880 and then Balmain Academy, and St David's College. In 1893 it was named Glendenning by a Mr J. H. Wise, an estate agent who purchased the property. The old home has known many occupants including medical practitioners and it is now the Manor House Restaurant. There are French doors to the veranda and a Victorian stone fountain in the garden.

Turn right into Birchgrove Road

Birchgrove was named by John Birch who was Paymaster of Governor Lachlan Macquarie's 73rd Highland Regiment. In 1810, soon after arriving in New South Wales, Birch bought thirty acres (12 ha.) of land originally granted to one George Whitfield in 1796. He named the property Birch Grove and built a two-storeyed stone house. It was a simple Georgian house, with an external kitchen (because of the danger of fires), and here Birch resided until 1814. In 1813 Governor Macquarie requested Lord Bathurst, Secretary of War and the Colonies, to recall the 73rd Regiment, complaining of their gross irregularity of behaviour and an alarming degree of licentiousness. The War Office had already decided to recall the regiment, and so John Birch was obliged to sail from Sydney in 1814. He sold Birch Grove House to a Mr R. W. Loane. Perhaps he missed his Sydney house; he certainly missed the governor's dinners for, in 1815, from Colombo he wrote to Captain Piper

> Sir Robert and Lady Brownrigg keep it up pretty well on the ball scale but not so much as in those pleasant family dinners which the General and Mrs Macquarie used to make every one so happy in.

Birch Grove House was demolished in 1967.

At the corner of Birchgrove Road is a former shop with the old advertising signs still visible. At no. 2 there is a veranda over a walkway, and old hitching posts left from the days when the horse was the chief means of transport. No. 11, the Anchorage, is a single-storeyed sandstone cottage and nos 21-27 are 1880s brick terraces.

Do not miss the Riverview Hotel. Its architecture is unusual and its most famous publican was Australia's greatest swimmer, Dawn Fraser, Balmain's favourite daughter. Dawn Fraser began her swimming career in Balmain's harbour pool and won more Olympic medals than any other Australian, proving herself to be one of the greatest swimmers of all time. Of eight Olympic medals, four were gold, and she won the 100 metres freestyle event in the games of 1956, 1960, and 1964. She set forty world swimming records and collected six Commonwealth Games medals. Dawn Fraser was awarded an MBE in 1967 and was the first Australian to be featured at the Swimming Hall of Fame at Fort Lauderdale, Florida, USA, in 1972. She served for one term as State member for Balmain in the New South Wales Parliament.

No. 33 Birchgrove Road has its original wooden tile roof. It is a two-storeyed timber and painted sandstone sea captain's house, complete with flagpole and yardarm at the front. The peninsula's history is scattered with the names of the sea captains who settled in the area.

Captain William Adam bought land in the Gilchrist subdivision (*see* p. 12). Captain John Roach was Master of *Prince George*, a revenue cutter, when he purchased land. Captain Frederick Trouton came to Balmain in 1866 and stayed thirty years. He went to sea at seventeen as a midshipman and had an adventurous career until he became a master at twenty-five. The goldrush brought him to the goldfields of Victoria in 1852 and in later years he became the general manager of the Australasian Steam Navigation Co. He died at Cliffdale House, Looke's Avenue, Balmain, in 1896 and is remembered in Trouton Street.

Turn left into Punch Street and left into Fitzroy Avenue for Elkington Park

Pass nos 2-12 Fitzroy Avenue, a row of six three-storeyed brick terraces, and into Elkington Park. Here is the Dawn Fraser Swimming Pool, originally built in 1883 as tidal baths, and birthplace of the Balmain Swimming Club. In 1964 the pool was renamed in tribute to the great swimmer. The Balmain/Birchgrove area did not always have pool facilities. The *Balmain Almanac* of 1878 states:

> There have been some public meetings — for baths, against abattoirs, or sanitary movements, & C. — at some of which a large amount of bunkum was used, but on the whole common sense prevailed. At one of the meetings held about the baths, one venerable father of the Council exclaimed 'Baths! — what do we want with baths? — I never bathe!' Happy man. What a bedfellow. On another occasion a gentleman said 'Give us roads — we can dispense with baths.' Balmain is almost surrounded by water and

yet without a public bathing-place. There was one place fixed upon for a bath, but at the extreme end of Balmain, and the thing fell through.

Because there were no public baths, accidents occurred:

> On one occasion a local boy Thomas Alfred Burless, about twelve years, went into the waters at the foot of Darling Street. He had not been in the water long when his companions heard his cries. A large shark had attacked. A Mr. Ridley dived into the water to save the boy. The muscles and flesh of his right leg had been torn off. He was taken to the Infirmary and the leg amputated.

The *Almanac* assured readers the boy was making a recovery but pointed out the attack was a warning to others not to bathe in the harbour.

From Elkington Park there are views of Snapper, Spectacle, and Cockatoo islands, Birkenhead Point and the Iron Cove and Gladesville bridges.

Snapper Island is the closest and during its history has been known as Flea, Mosquito, Cat, and Rat Island. The island has had a long association with the volunteer sea cadets movement. Commander Leonard Forsyth established the base to train sea cadets in 1932; he was also responsible for founding the island's museum which contains many relics of Australia's naval history. Commander Forsyth died in 1983 when he was ninety-one years of age. In 1987 the Royal Australian Navy ordered the cadets to leave Snapper Island because of the risk from ammunition barges moored nearby.

To the north west of Snapper Island lies Spectacle Island, originally called Dawes Island. In 1863 the island was chosen for use as a powder magazine, and stone buildings were designed by architect James Barnet. The old buildings carry the date 1865 and were built by convicts from Cockatoo Island. The island also has many relics of naval importance which will be transferred to Sydney's new Maritime Museum at Darling Harbour.

The large island is Cockatoo, so named because of the white cockatoos that nested in its trees. The island was a sandstone hillock of some forty acres (16.2 ha.) when in 1839 Governor Gipps had grain stores hewn out of the rock. The convicts laboured on the island and it was a gaol for hardened prisoners. Eighteen grain stores were cut from the rock, each twenty feet (6.1 m.) deep, but in 1841 the English government ordered the silos to be closed. England was adopting a free market policy at the time.

Governor FitzRoy laid the foundation of the dry dock in 1853 and again prisoners were used on a Sydney island to do the manual work. The dock was not completed until 1857 and the first vessel to enter the dry dock was HMS *Herald*.

Cockatoo Island has a grim history and was used as a place of confinement for male prisoners until 1871. It then served as a girls' reformatory and women's prison until 1909. After the departure of the women the island was called Biloela in an attempt to remove the stigma of the women's prison from the island.

The male prisoners had commenced the first dock on the island but a second dock was completed in 1890. The New South Wales Government ran the docks and shipyards from 1908 until 1913 when the Commonwealth Government took control of Cockatoo Island as a naval dockyard. In 1933 the island was leased to Cockatoo Docks and Engineering Co. Ltd. During World War II 3,600 men were employed on naval shipbuilding and repair work but when Captain Cook Dock was built at Garden Island in 1945 much of the work transferred to that island.

The vehicular vessels for the Bass Strait crossing to Tasmania were built at Cockatoo Island. (For further information about Cockatoo and other islands see Simon Davies *The Islands of Sydney Harbour* Hale & Iremonger.)

In the distance is the Iron Cove Bridge spanning the inlet of Iron Cove which separates Balmain and Drummoyne. The bridge dates from the 1950s and replaced the wrought-iron lattice bridge of 1884. Birkenhead Point, before the bridge, is a shopping complex created on the foundations of the old Dunlop rubber factory.

From Elkington Park return up Glassop Street and turn left into Birchgrove Road

At nos 41-43 Glassop Street is Millie Villas. There is a tunnel between the backyards. Look at the tiled panels to the windows, the parapet, and pediment.

Wander along Birchgrove Road

Note the architecture: no. 54 Lilywell has an end terrace extending over the pavement; nos 59 and 57 are single-storeyed semis, notice the point on nos 63 and 61; nos 68-66 are two-storeyed semi-detached houses with decorative iron lace; no. 75 is named St Kilda and is the same design as no. 77, though the latter has an attic and timber barge boards.

Houses from the period 1840 to 1870 often have touches of Gothic styling, sometimes tall chimneys, casement windows, and cusped timber barge boards. The roofs are high pitched and building materials were used in a decorative manner.

On the left is Birchgrove Public School

Birchgrove Public School opened on 9 March 1885 with 241 pupils. With a growing population, additions were made in 1887. By 1889 school guards had to be placed at the windows in Birchgrove Road to warn the staff of larrikins throwing stones at the building. In the early 1900s a two-storeyed brick building was added to the school which still serves the pupils of Birchgrove. Norman Lindsay, artist and author, describing his school in Victoria in the 1890s, spoke of 'the fusty odour of crowded classrooms' and 'the discomfort of hard seats'. There is no reason to assume New South Wales school differed from their Victorian counterparts. Old photographs show austere classrooms where the use of the cane was common. In 1901 a teachers' conference in Sydney complained teaching was mechanical and stereotyped, the examination system oppressive, higher education limited, and teachers inadequately trained and paid.

Birchgrove Public School also had to contend with a coalmine operating next to the school from the 1890s to the 1930s.

The Balmain Coalmine began in 1897 to tap the resources of the Bulli coal seam running 3,000 feet (914 m.) below Sydney harbour. The deposits of coal, among the largest in the world, are in a saucer-shaped basin covering an area north of the Hunter River Valley, west to Lithgow, and south to Wollongong. The centre of the basin is beneath Sydney harbour and was known as early as 1847. Tests were made at Cremorne in 1893 and in 1903 a company, Sydney Harbour Collieries Co., was formed to work the coal seams. Balmain was chosen as the best site for a shaft. Two shafts were finally sunk, each eighteen inches (46 cm.) in diameter and named the Birthday and Jubilee shafts, for Queen Victoria's birthday and Diamond Jubilee.

There was plenty of good quality coal and the first coal was brought up and sold in June 1902 — eighty to 100 tons per week. Coal was extracted for fourteen years but the mine closed in 1917 because it was no longer economic. As well as the financial problems, water seepage caused technical difficulties. Despite efforts to re-open the mine the Department of Mines, refused to renew the lease in 1930. Conditions at the mine had been harsh: the men at times worked in temperatures of 100°F (38°C).

Between 1932 and 1937 natural gas was extracted from the mines. Finally in 1945 the old mines were sealed but three men were killed during this operation when escaping gas led to an explosion.

Birchgrove School was probably glad to see its end.

Turn right into Macquarie Terrace and left into Thomas Street

Thomas Street is a narrow street lined with timber dwellings. Some have quaint names such as 'House Next Door' and 'House Opposite'.

Turn left into Spring Street

Walk past the former corner shop with its advertising signs and proceed to the site Jubilee Engineering Co. Pty Ltd and Howard Smith's in Water

Street. The Howard Smith site was to become a housing development but building has yet to commence. This area was part of the Balmain colliery and site of the Jubilee and Birthday shafts.

Continue along Birchgrove Road

St John's the Evangelist dates from 1882. The architect was E. H. Buchanan, who also designed Balmain Town Hall. The foundation stone was laid by Dean Cowper on 4 February 1882, and the building was intended as a school/church. It cost £1,200 to build and has had various additions made to it. The building originally extended only as far as the present chancel steps with vestries and classrooms behind the altar. The church has had a number of interesting rectors including the Reverend Raymond King, whose brother Copeland King was one of the founders of the Diocese of New Guinea. Reverend King was succeeded by E. J. Sturdee, brother of Admiral Sturdee. The Reverend W. A. Charlton, appointed in 1889, was a great sportsman and helped establish the Birchgrove Oval. He was followed by Reverend W. T. Cakebread. Both Cakebread and Charlton were chaplains to the reformatory ship *Sobraon*.

The *Sobraon* was a popular sailing ship on the England-Australia run in the 1860s. Built of solid teak with iron beams and frames, in the 1890s it became a ship for juvenile delinquents and was a familiar sight moored at West Circular Quay. Sydney boys were warned if they did not behave they would 'end up on the *Sobraon*'. It was later re-named *Tingira* and became a naval training ship before being dismantled at Berry's Bay in 1936.

Do not miss the lion's head fountain in the churchyard. It came from the first Bank of New South Wales building in Sydney, which stood in Macquarie Place not far from Mary Reibey's cottage where the bank had been founded in 1817.

From Birchgrove Road turn left into Cove Street

Federation houses are characterised by red bricks, red-tiled roofs, carved timber trimmings, and often possess wide verandas with turned wooden posts. Nos 5-15 are Federation-style terraces. Contrast them with nos 6 and 8, which are single-storeyed sandstone cottages. Nos 10 and 12 are double-storeyed brick, and nos 51 and 53 timber cottages with fine iron lace work.

Enjoy the vista from Cove Street before turning into Louisa Road

Didier Numa Joubert, French pioneer of Hunter's Hill, owned land in Birchgrove and he named Louisa Road for a family member. Joubert, from Angoulême, arrived in Sydney in 1837 as a wine and spirit merchant. He subdivided his Birchgrove land, some of which was originally part of the Birch Grove House estate.

Louisa Road leads down to Long Nose Point and has many fine homes. First note no. 12 Keda *c.*1878, and no. 14 Lenardville *c.*1876 is particularly interesting. It has been restored and renamed Vidette. There is a well in the left-hand corner of the front garden. It is seventeen feet (5.2 m.) in depth and is still fed by a natural spring. Logan Brae, no. 24, was built about 1907 and the Royal Australian Institute of Architects, NSW Chapter, says 'its ornateness is quite exceptional for Balmain, even though the area was experiencing a building boom at the time'. Its style is art nouveau and there is tuck-pointed brickwork with elaborate veranda timber. Note also the tower and wooden walk. No. 28 has wide weatherboards, iron posts, and an attractive garden.

The Anchorage, no. 44, is a district landmark. It was built in the late 1890s and is five-storeys high at the back. It has an attic storey and cast-iron widow's walk. The land runs down to the waterfront. It was first known as Fitzroy House and was built by John Gibson who owned an engineering business, J. Gibson & Son. The property was purchased in 1920 by the Driscolls, sawmillers and timber merchants. The waterfront on the Parramatta River would have been an advantage for ships would have been moored nearby.

Opposite is no. 63 Glenree, which has trim iron work and stained-glass panes to the windows. No. 67 was the site of Birch Grove House, which was replaced by flats in 1967. There had originally been a farm and orchard, and the house had survived from the early 1800s but was demolished despite public protests.

Different aspects of the houses in Louisa Road will catch your attention — a tiny porch, a latticed area, or a wooden barge board. There are workmen busy at some premises, restoring or re-building and it is evident the old houses are still part of a functioning community. There is a path to and view over Birchgrove Park and oval, and Snails Bay. The park was reclaimed in 1880 after filling with rubble from the Balmain colliery. The oval was once the home of the Balmain 'Tigers', who now play at Leichhardt Oval. Balmainites urge their team to victory with cries of 'Up the Tigers'.

There is a fine stone house at no. 76 Louisa Road. This land was part of the Birch Grove House estate but, as the carved stone at the top of the house declares, this residence was built by J. Lord in 1881. John Lord had an orchard and sawmill at Kurrajong and built himself a strong stone house in Louisa Road. The stone was quarried on the site, but Lord had little chance to enjoy his new home as he died soon after its completion. Note the little iron lace Juliet balcony and the unusual zinc barge boards. There is a strong stone and railing fence to the property.

No. 85 is a gracious old home with grounds sweeping down to Snails Bay. Note the curlicue barge board and tessellated tiled veranda. It is named Geierstein and was built by Alexander William Cormack. Cormack was born in 1837. His parents were John Hunter Cormack, a cooper, and his wife Maria Fulloon. Maria's mother, Elizabeth Fulloon, had been Superintendent and Matron of the Female Factory at Parramatta in the 1820s. Cormack owned several cooperages in the 1880-1890s and died at Geierstein in 1909. Nearby new town houses use timber in the old style. No. 109A even has a Juliet balcony. Raywell, no. 144, has had a checkered career. It stands back from the pavement on a slope of ground and is enclosed by a stone and railing fence. It has pretty veranda posts but the tiled roof does little for its style. It was built in 1883 by Duncan Smith and sold two years' later to a produce salesman named Ainsworth. He in turn sold it to a Miss Rachel Wells and she named the house Raywell. Additions were made in 1891, and the house has since suffered the indignity of being subdivided into flats. It was in a deplorable state in the 1970s but restoration was undertaken in the 1980s.

It is a pleasant stroll down to Long Nose Point where the homes have wonderful water views.

At the end of Louisa Road stands a 1920s three-storeyed sandstone and timber residence. Once the house was headquarters for a motor bike gang but is now a family home and has been restored to its original state.

Long Nose Point is the western extremity of Port Jackson and is a mile-and-a-half (2.4 km.) from the harbour bridge. The point is a sandspit and marks the mouth of the Parramatta River. On the opposite side of the point is Greenwich, and the confluence of the river and harbour is known as the Watersmeet.

Long Nose Point Reserve is a tranquil area from which to view the passing maritime scene and won its architect, Bruce McKenzie, an award in 1979. The land was donated to the council authorities by the firm, Morrison & Sinclair, boatbuilders and repairers. The point attracts local fishermen who in February haul in fine blackfish. The casuarina trees soften the contours of the sandstone outcrops, and quiet paths bid lazy wanderings in this superb setting.

A ferry to the city leaves Long Nose Point at hourly intervals, but there is still much of Birchgrove to be discovered and the walk continues back up Louisa Road to Deloitte Avenue

Deloitte Avenue crosses Birchgrove Park and overlooks Snails Bay. The bay may hav been named for molluscs known as seasnails (genus *Aplysia*). John Birch is said to have moored his ship in the bay and in a later era ships from New Zealand, New Guinea, and North America unloaded their cargo of log timber at dolphins. The dolphins were clusters of

Walk 3: Birchgrove

Birchgrove Park and Snails Bay, c.1891 (ML)

buttressed piles and the logs were towed to timberyards and sawmills. There are old wooden structures on the shoreline, deep placid waters, and the fascination of the sea and ships.

A street at the rear of the park is Rose Street and the sculptor Achille Simonetti worked here on his statue of Governor Arthur Phillip which graces the Sydney Royal Botanic Gardens. Simonetti had been persuaded to settle in Balmain by Premier Sir Henry Parkes.

Leave Birchgrove Park, cross Grove Street and enter Wharf Road opposite

Wharf Road has some very old houses — nos 15 and 15A are from the 1830s and have stone floors quarried on the site. Additions were made in the 1850s and attics added in the 1880s. No. 15 has been restored but the original cedar joinery, Huon pine ceilings, and Kauri floorboards remain. Nos 33 and 35, named Simla and Oneida, were the homes of the Deloitte family for many years. Captain William Deloitte sailed from England with his seven sons in 1837. He lived briefly at Birch Grove House. (In 1828 architect John Verge arrived in Sydney on the brig *Clarkestone* captained by Deloitte.) The cottages were built in the 1840s from stone quarried at Pyrmont. The roofs are of Welsh slate. Both were later bought by Lavinia Deloitte Vernon, the Comtesse de Vilme-Hautemont. She was a personality in the musical comedy world.

Wharf Street is quiet with old stone steps leading to some houses and iron railings festooned with creepers. No. 34 is an attractive stone residence with shuttered windows and attractive gate posts and white painted gate. No. 39, Ravens Court, on the water side of the street, is an elaborate Italianate house. This style first became popular in the 1840s but blossomed in the period 1870 to 1890. Ravens Court has rare iron gate posts and the house an unusual front window; the walls are decorated with garlands of flowers. There is a tower with views to the harbour, a tiled entrance porch, decorative barge boards and a pretty bowed cast-iron veranda facing the garden and water. For many years Ravens Court was a boarding house but has now been restored.

Wansworth, no. 28, is also old, *c.*1840. Pink and cream frangipani flower in the garden. No. 26 has a flight of steps

Iron posts and gate at the entrance to Ravens Court

leading to the stone house. No. 22 is triple-storeyed of timber and brick and decorated with green cast-iron. No. 25A has a little porch addition on its side wall and steps lead to the harbour garden. Normanton, no. 21, is painted yellow ochre and is close to Lemm Street.

Lemm Street crosses Wharf Road

Cross Lemm Street, named for local builder Frederick Lemm, and look at the gabled house on the corner.

Walk to 8 Wharf Road

It is only a few footsteps to no. 8 Wharf Road, which has cut sandstone ledges to the windows, a tiled veranda with iron railing. Note the old timber garage opposite with a turned point to the gable and a scarlet grape-vine hanging over the opened doorway.

Return to Lemm Street and turn left into Ballast Point Road

No. 29A was owned by the builder Frederick Lemm. Note the joist with hook for loading building materials. Tyne Villas, 1886, stand at nos 25-7 and are semi-detached houses. No. 23 is named Bremer House.

Retrace your steps to the corner of Yeend Street and go uphill

The old Australian National Line area has been cleared for government housing and there is a view to Balmain with St Augustine's Church dominating the skyline. Ellerslie Terrace, nos 35-47, dates from 1881. There are barley sugar posts at the windows. One resident was Quartin Deloitte, who was associated with the Balmain Regatta (*see* p. 11). The scullers were heroes at the regattas and a native-born Australian, Ted Trickett, beat the English champion sculler J. H. Sadler in 1876, in a contest on the River Thames in England. The first Australian to win the title, Trickett was a giant of a man, six-feet three-and-a-half inches (192 cm.) tall and weighed thirteen stone (83 kg). When he returned to Sydney 25,000 people welcomed him at Circular Quay. He lost the title four years later to a Canadian. A public subscription raised £850 to buy Trickett his own hotel

Other residents of Ellerslie in 1881 were George Kebblewhite, chemist; David Watson, gentleman, and a Mrs Campbell.

Continue along Ballast Point Road

At the corner of Short Street there is an acutely angled building jutting onto the pavement. At the intersection of Dock Road the building on the left-hand side has a decorated valance board. Nos 46-56 Ballast Point

Road is named Yeroulbin, meaning swift-flowing waters, and was built at the turn of the century by a local alderman named Turner. Unfortunately it is in need of restoration as it shows its age. Clifton Villa, no. 73, is one of the most photographed Birchgrove residences. The land was purchased by a J. Perkins at the time T. S. Mort was busy buying land for his dock. Built in the 1860s in the Gothic Revival style, the villa has eight bedrooms, including four in the attic. In 1871 a steamboat proprietor, John Manning, resided here and in 1876 Mrs Helen Chilcott opened a Young Ladies' School, where pupils were taught elocution, deportment, and home skills. It has had numerous owners since those days and in the 1970s was an art school conducted by Charles Blackman. Look at the handmade barge boards and the rusticated window-dressings. The grand mansion also has an Edwardian ballroom and more modern additions including a swimming pool.

Walk down steps at Harris Park Reserve to Grove Street and turn left

A group of terraces at nos 56-69 are rather shabby but are unusual as some have outside steps leading to the first-floor level. A new terrace has been built alongside the old. Further along Grove Street, which was extended from Birchgrove Road in 1882 when land was resumed at Snails Bay, are a row of stone terraces, nos 39-55. There are eight terrace houses built in 1886 and named Gladstone Terrace. At the end of the row is the old corner shop with a room overhanging the pavement.

Turn left into Cameron Street

A corner turn leads to another era for here is the old Birchgrove village. Opposite the Grove Street corner shop is a two-storeyed building with a veranda. It is easy to identify the old two-storeyed timber shops and look at the row of little wooden cottages in Gipps Street. The village retains the atmosphere of an earlier age. The streets evoke memories of the horse-drawn carts of the baker and 'milko' or the 'ice man' rushing by with his block of dripping ice.

Walk far enough down Cameron Street, named for Ewen Cameron, a founder of the Municipality of Balmain, to glimpse the Sir William Wallace Hotel at the corner of Spring and Short streets. This is a two-storeyed corner hotel with a panelled lace iron upper veranda and wooden posts. The old hotel was used in the Australian film *Caddie*.

Walk along Rowntree Street to Macquarie Terrace

In Macquarie Terrace shaded by leafy trees is the Rowntree Monument (*see* p. 43 and Profile). The craggy stone pedestal and column bears a cross and anchor which came from Rowntree's grave in Balmain Cemetery* and carries the following inscription:

> ERECTED TO THE MEMORY OF
> THOMAS STEPHENSON ROWNTREE
> BY THE BALMAIN MUNICIPAL COUNCIL
> ALD. M.D. CASHMAN J.P. MAYOR 1941.
>
> BORN AT SUNDERLAND DURHAM 6TH JULY 1818
> DIED NORTHUMBERLAND HOUSE BALMAIN
> 17TH DECEMBER 1902 IN HIS 85TH YEAR
>
> SWIFT TO ITS CLOSE EBBS OUT LIFE'S LITTLE DAY
> EARTHS JOYS GROW DIM ITS GLORIES PASS AWAY
> CHANGE AND DECAY IN ALL AROUND I SEE
> O THOU WHO CHANGEST NOT ABIDE WITH ME

It is fitting to end the Birchgrove walk at a memorial dedicated to an early pioneer of the district. Much of the old district that Rowntree knew has survived to give pleasure to a later generation.

It is a short walk up Rowntree Street to Darling Street.

---

* The site of Old Balmain Cemetery is the Pioneer Park in Norton Street, Leichhardt. With the destruction of the cemetery in the 1940s a rich record of pioneers of New South Wales was lost.

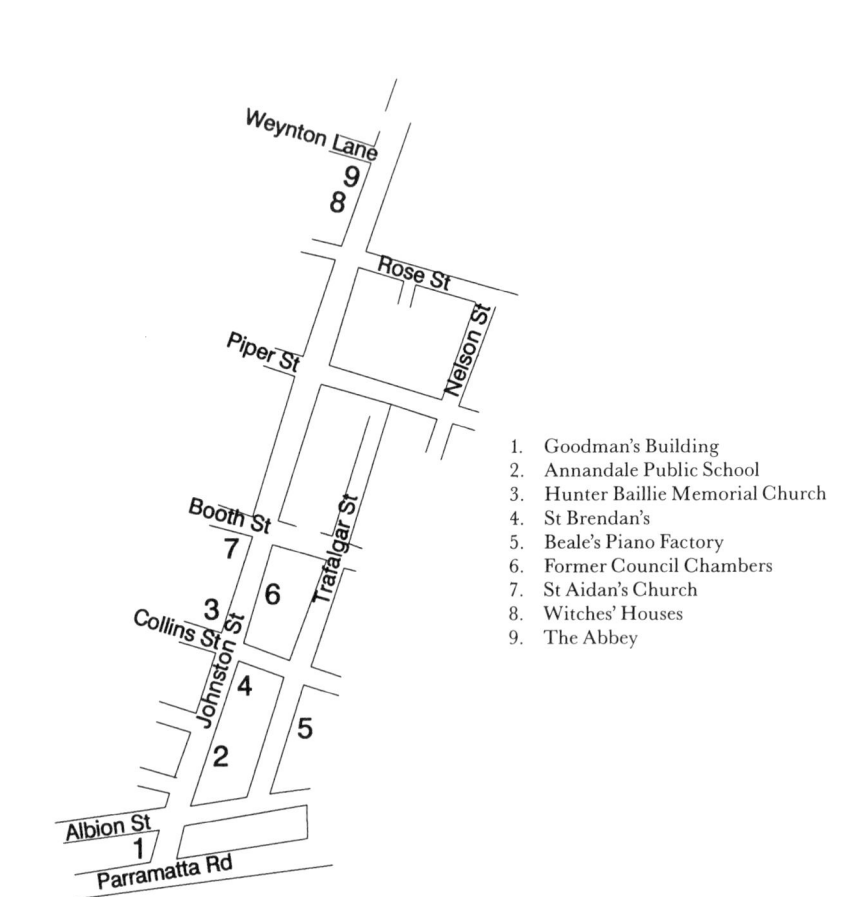

1. Goodman's Building
2. Annandale Public School
3. Hunter Baillie Memorial Church
4. St Brendan's
5. Beale's Piano Factory
6. Former Council Chambers
7. St Aidan's Church
8. Witches' Houses
9. The Abbey

Walk No. 4

Annandale

George Johnston came to Botany Bay with the Marines who accompanied the First Fleet. He was one of the first landholders from the officer class in the colony and received 100 acres (40.5 ha.) at Petersham Hill on the Parramatta Road in February 1793. Johnston named his estate Annandale for his birthplace in Dumfries-shire, Scotland.

Johnston had joined the Marines in 1776 as a Second Lieutenant and had seen service in New York, Halifax, and the East Indies. In 1786 he transferred to the Marine detachment selected for Botany Bay and sailed on the *Lady Penrhyn*. On the transport Johnston was in charge of discipline and preventing disorder among the women convicts. It was on the *Lady Penrhyn* that he noticed dark-eyed, black-haired convict Esther Abrahams who was twenty years of age. She was Jewish, a milliner, and had appeared at the Old Bailey in August 1786 charged with attempting to steal twenty four yards (22 m.) of black silk lace. For this offence Esther was sentenced to seven years' transportation and lodged in Newgate Prison pending her departure for Botany Bay. In March 1787 she gave birth to a daughter, Roseanna, and the baby sailed with her on the *Lady Penrhyn*. Perhaps it was her dark beauty that attracted the fair Scots officer, for Esther was soon under the protection of Lieutenant Johnston.

In March 1790 their first son, George, was baptised in Sydney Town only two days before they sailed for Norfolk Island. Johnston returned to Port Jackson in February 1791 because of ill health, and Esther Abrahams returned in May the same year. Their second son, Robert, was born on 9 March 1792. That year Governor Phillip gave Johnston command of a company of the New South Wales Corps raised from the Marines. In 1786 he was aide-de-camp to Governor Hunter.

At his Annandale property Johnston planted Norfolk Island pines,

A watercolour of Annandale House painted in 1877 by Samuel Elyard (ML)

believed to be the first brought to Australia. By 1805 he had been granted 2,000 acres (810 ha.) at Cabramatta, and was given a further 1,500 acres (608 ha.) at Illawarra in 1817.

Annandale House was a comfortable home, and Johnston's and Esther Abrahams' third son, David, was born there in 1800. The estate had its own bakery, smithy, slaughterhouse, butchery, stores, vineyard, and orangerie. In 1800 Johnston was placed under arrest by his commanding officer, William Paterson, for 'paying spirits to a serjeant as part of his pay at an improper price, contempt, and disobedience of orders'. Johnston objected to a colonial court-martial and Governor Hunter sent him to London where his trial was aborted because of the difficulty caused by the witnesses being in New South Wales. During his absence Esther ran the Annandale estate until Johnston returned in 1802.

Johnston led the troops who suppressed the Irish convicts in the rising at Vinegar Hill (Rouse Hill) in 1804. He also had command of the New South Wales Corps on 26 January 1808 when the troops arrested Governor William Bligh in the Rum Rebellion. As a result of that action Johnston was again sent to England in 1809 and faced a court-martial in 1811. He was found guilty of mutiny but was fortunate only to be cashiered. In 1813 Johnston returned to Sydney as a private citizen and lived quietly at his Annandale estate. Governor Macquarie frowned upon the many liaisons in the colony, and Johnston married Esther Abrahams on 12 November 1814 at Concord. The service was conducted by the Reverend Samuel Marsden, and Roseanna, Esther's baby on the *Lady Penrhyn*, was a witness. Roseanna was wife to Isaac Nichols, the first postmaster in the colony.

The Johnstons had three sons and three daughters but the first born, George, was killed at Camden in a riding accident on 19 February 1820.

Three years later his father, George Johnston, died on 5 January 1823 and was buried at Annandale in a vault designed by Francis Greenway. The estate passed to Esther and their son Robert. In 1829 Esther considered morgaging the property and returning to England. A family rift developed and Robert endeavoured to have his mother declared insane. The court proceedings of March 1829 found that Esther drank, quarrelled frequently with her children, and that her habits, if not insane, were eccentric. The jury found her 'insane but having lucid moments'. Trustees were appointed to the estate and Esther moved to her younger son David's property at George's Hall, Bankstown. She died there on 26 August 1846 aged seventy-five years and was buried with her husband George in the family vault. Their remains are now interred in the Johnston family vault at Waverley Cemetery, Sydney.

Robert Johnston joined the Navy in 1807. He and Phillip Parker King were the first two Australians to join the British Navy. Johnston returned to New South Wales in 1816 and took part in various explorations in the colony. He later engaged in farming but by 1876 he had begun to subdivide North Annandale. He offered blocks of land sixty-six feet (20 m.) by 190 feet (58 m.) in an area bounded by Parramatta Road, Johnston, Booth, and Nelson streets and the suburb of Annandale grew to maturity in the 1880s.

Our walk commences at the corner of Parramatta Road and Johnston Street

Johnston's original 100 acre (40.5 ha.) grant of 1793 was enlarged with a further eighteen acres (7.3 ha.) and another twenty-two acres (9 ha.) in 1794. These grants were on the southern side of Parramatta Road where the present-day suburb of Stanmore now exists. In 1799 Johnston received 290 acres (117.5 ha.) on the north side of Parramatta Road and this land stretched to Rozelle Bay. Old Annandale House was demolished *c.*1903 and one can only imagine the home with its long driveway of Norfolk pines. The house stood on the southern side of Parramatta Road and an old gate house, dating from the 1870s, still stands in the backyard of 96-98 Corunna Road, Stanmore. It originally stood on Parramatta Road.

Goodman's Building

On the corner of Parramatta Road (western side) stands Goodman's Building, a grand commercial building, so close to the city. It has a wide veranda and balcony, typical of country town architecture. There is a balustrade which hides the corrugated iron roof of the building, which was erected between 1893 and 1912.

Walter Goodman was a local shoe merchant and entrepreneur. The architect of the emporium was Joseph Sheerin of Sheerin and Hennessy, architects of St Patrick's College at Manly.

The Department of Environment and Planning has restored this example of 'Boom Style' architecture. The Department of Main Roads objected to the veranda posts on the Parramatta Road side of the building and they have been removed and the bullnose iron modified. The posts remain on the Johnston Street side.

Annandale was to be a model suburb. Ferdinand Reuss Jnr, architect and surveyor, won a competition for the planned suburb. He designed a spacious and regular grid-road pattern, unique in Sydney, with streets sixty-six feet (20 m.) wide. Annandale Street is eighty feet (24.4 m.) and Johnston Street 100 feet (30.5 m.). It was intended to be the finest street in the colony.

Walk down Johnston Street

Note no. 1 on the right-hand side with its cast-iron columns and the urn at the top of the house. Most of the buildings in Annandale were built between 1880 and 1910 and are late Victorian in style, but there are some Art Nouveau and Federation examples. To the left, the houses are Federation red brick with iron railings and tiled steps and verandas.

Turn left into Albion Street

There has been little redevelopment in Annandale and it has been estimated ninety-five per cent of the houses are original. Note the houses with wooden barge boards in Albion Street and the two-storeyed homes decorated with cast-iron. There are narrow terraces also decorated with barge boards. It was part of Sydney's Gothic fetish to suspend wooden barge boards from the gables.

Holy Family Convent

The convent was originally Macquarie Lodge, a house on the Johnston estate. It was built between 1830 and 1850, although it has since been modernised. In the 1890s it became a refuge for unmarried mothers, founded by George Lewis, who was appalled at the number of illegitimate babies being murdered. Even without resorting to murder, infant mortality in densely settled suburbs was sometimes as high as forty-six per cent of all births. In the 1880s there were 133 deaths per 1,000 births. Mothers died in childbirth, and typhoid, diarrhoea, diptheria, croup, and 'infant convulsions' claimed the children.

Return to Johnston Street

Note the view of the University of Sydney from Albion Street. A flag flies from the tower of Edmund Blacket's Great Hall. On the corner of Johnston and Albion streets is Arthurleigh, one of the great houses of the 1880s.

Note nos 24, 26, 28, 30, and 32. Most are decorated with tiles of attractive floral designs. No. 24 has been restored. It was built by a Mr O'Ridge, a local stonemason. His son did the carpentry work at no. 24. The land was purchased from the Johnston estate and the property remained in the O'Ridge family until 1964. Monsignor O'Brien, head of St John's College, University of Sydney, was the next resident.

The roofs of the houses were mostly slate as shingle roofs began to disappear by the 1860s. The slate was brought out as ballast by the wool clippers sailing to Australia. The houses are often decorated with plaster ornaments and cast iron. Australia followed British trends, and many terraces and houses still featured parapets originally designed to stop snow falling on passers-by.

Cross Johnston Street at traffic lights to Annandale Public School

In the grounds of the school are the gates of Annandale House. After the estate was broken up the gates for some years were left lying behind the Annandale Council Chambers. In 1974 a suggestion was made that they be re-erected in the grounds of Annandale Public School. A memorial plaque states:

> These gates once stood at the entrance to Annandale House on the south side of Parramatta Road west of Johnston Street. Annandale House was built about 1799 by Col. George Johnston on land granted to him on his arrival with the First Fleet.
> Annandale House was demolished in 1914. [There is a suggestion that this actually occurred c. 1903.]
> Re-erected on this site in 1977 by the Department of Education with the co-operation of the Council of the Municipality of Leichhardt and the Annandale Association.

Re-cross Johnston Street to no. 60

Clydebank is two-storeyed and decorated with tiles and iron work. Note houses in this row to the corner of Cross Street. Their names are inscribed on the stone gate posts.

Walk into Cross Street on the left

Note the little cottages with their high-pitched roofs and barge boards. Most of the old slate roofs have been replaced with red tiles which look clumsy and heavy on the houses. In the distance is the soaring spire of the Hunter Baillie Church.

Return to Johnston Street and note the corner house with its rear balcony. Nos 66, 68, and 70 are more fine examples of two-storeyed residences with plaster decoration and spear iron fences. The rich ornamentation of plaster and cast-iron is typical of the boom years of the 1880s, but by 1920 Ussher in *Australian Cities* was describing such work as 'a hideous legacy . . . florid and meaningless and in execrable taste'.

On the opposite side of the road notice nos 33, 35, 39, 41, 45-47. No. 30, Norton House, is built of red and white brick in a 'blood and bandages' style of brickwork. No. 35 has a fine tower and spire. Note the plasterwork, arches and tower of No. 39. No. 41 is of the same style. Nos 45-47 are now flats, but vestiges of the house remain in the lion's head over the entrance.

In the 1870s and 1880s a young professional man, such as a solicitor, might earn between £10 to £20 a week, three to six times as much as a skilled artisan. A single-storeyed villa with land cost approximately £1,500; £10 a week represented two or three years' salary. A man earning £20 a week in 1886 could aim for a two-storeyed home, almost a mansion, built at a cost of £2,500 to £3,000.

The classic style in the 1870s was popular and houses had parapets, motifs of garlands, and urns. The iron work was extravagant and patterns were often picked out in different colours. Paint scrapes of old iron has often revealed a taste for a muddy green or brown.

On the corner of Johnston Street and Collins Street stands the Hunter Baillie Memorial Presbyterian Church whose soaring spire is an Annandale landmark. John Hunter Baillie was born at Hamilton, Scotland, in 1818. When he was twenty-three he sailed for New South Wales and while in quarantine in Port Jackson he wrote to Dr John Dunmore Lang, the Presbyterian clergyman in Sydney, for advice. Lang was impressed by Baillie's letter and offered him a position as a sub-editor of the *Colonial Observer*, a newspaper Lang had established. Baillie lived with the Langs and met Dr Lang's sister-in-law, Helen Hay Mackie. The couple married in January 1844. Baillie later became a financial adviser to the Bank of New South Wales and was later appointed to the staff. He suffered constant poor health and a period in Victoria for the bank did nothing to improve his constitution. He travelled extensively for the bank in New South Wales and Victoria and the rigours of travelling may have hastened his death. It is possible he was suffering from tuberculosis or had an ulcer. He died on 25 March 1854. His widow, Helen Mackie Baillie, devoted her widowhood to philanthropy — she is remembered by a ward named in her honour at the Children's Hospital, Camperdown. She built the Hunter Baillie Church as a memorial to her long-deceased husband. The

The soaring spire of Hunter Baillie Memorial Church, Annandale

church was erected between 1886 and 1889 and the first minister was her nephew, the Reverend Peter Falconer Mackenzie, who had married Dr Lang's daughter, Isabella.

The church was designed by Cyril and Arthur Blacket, sons of architect Edmund Blacket. They first built the church hall, which was used for services and meetings while the church was being constructed. The foundations of the church were commenced in 1886 — the first design was for a Romanesque style but was rejected by Mrs Baillie. The second design with a spire 150 feet (46 m.) was accepted although the spire was later elongated to 182 feet (55.5 m.). The stone is Pyrmont sandstone and inside the church are pillars of Aberdeen granite with Victorian bluestone bases. A feature of the columns are the brass

Daisy-patterned iron at 11 Collins Street, Annandale

coronets with an alternate design of thistles and crosses which hid the original gas lights. The thistle motif is repeated in the gas light stands above the pulpit. The pulpit rails are carved in an oak leaf pattern and the pulpit is Oamaru stone from New Zealand. The stained-glass windows have a geometric pattern and the church is small inside. Possibly the organ crowds the church and the original plans were for a smaller organ. The organ is the work of Hill & Son of London and was installed a few years after the church was completed. It is said to be one of the best Hill organs in Sydney.

The full history of the church is covered in an excellent booklet entitled *Hunter Baillie* by Alan Roberts and Elizabeth Malcolm, which is available at the church.

Diagonally across the road from the Hunter Baillie Church is St Brendan's Roman Catholic Church. Its foundation stone was laid on 11 August 1912 by Archbishop Kelly, Archbishop of Sydney from 1911 until 1940. Roman Catholics became active in Annandale in the late 1890s. In the church English oak was used in the altar and interior fittings. The logs were unloaded and floated in Rozelle Bay before being hauled up Johnston Street by drays. In 1986 the church was restored, but as English oak was unavailable, North American oak was used. The timbers are well matched, resulting in a successful restoration. The new baptismal font is unusual. Leading to the pool is a cord of chains which symbolises a fountain of living water. The anchor patterned in mosaic in the pool is the symbol of hope. The upper gallery with its raked floor has also been restored.

Both the Hunter Baillie and St Brendan's are active churches that combine with the congregation of St Aidan's Church of England and the

Uniting Church of Annandale in an ecumenical gathering each year. The church workers erect stalls in the grounds of the Hunter Baillie and the combined fete attracts a large crowd. Sadly it is a sign of our age that few churches can remain unlocked at all times because of vandalism.

Next to the church in Collins Street is the former St Joseph's Convent. This was the original home in this area, and the building has been restored. The original iron lace, a rare daisy pattern, could not replaced. Another cast-iron railing has been used. The black and white veranda tiles had been vandalised and were beyond restoration. Note the long windows.

Opposite are two semi-detached houses. They are large double-storeyed homes with towers. No. 11, with its plaster urns, has daisy-patterned cast-iron similar to that which originally graced the convent. Next door at no. 9 is Agincourt, another two-storeyed home with cast-iron trim.

On the corner of Collins and Trafalgar streets (no. 3 Collins) is Wootten, home of a timber merchant, Sir Allen Taylor. Taylor was Mayor of Annandale from 1897 to 1902, and later Lord Mayor of Sydney. He also became a member of the State Parliament of New South Wales.

Turn right into Trafalgar Street

Note the old corner shops and proceed to no. 49. This Italianate bungalow is well proportioned and its most interesting feature is the lovely English Minton tiles. The various tile panels depict English birds and country scenes. The Minton factory was founded in 1793 at Stoke-on-Trent by Thomas Minton. He started making bone china in the 1820s and created the famous willow pattern. Nos 43-47 Trafalgar Street was Beale's Piano Factory. This huge building was opened in 1902 by Sir Edmund Barton, Australia's first prime minister, and a brass plate at the entrance records the fact.

The building is still very impressive and is surrounded by a fine wrought-iron and cast-iron fence. The first piano in Australia arrived with the First Fleet and was brought out by Dr Worgan. He later gave the instrument to Mrs Elizabeth Macarthur. Pianos played an important part in Victorian life, residing in drawing rooms furnished with cretonne-covered sofas, a davenport or ottoman, perhaps some easy chairs, a heavy sideboard, and a mantelpiece cluttered with bric-a-brac and mementos. That the piano was still an important part of social life at the turn of the century is evidenced by the size of Beale's factory. The factory closed during World War II

The Beale Piano Factory, Annandale, which opened in 1902

No. 41 Trafalgar Street is Edwinville

This house was designed by George Allen Mansfield of Glebe and built for the Sheriffs, a family of stonemasons from Scotland. Note the bow windows and the elaborate bands of stone carvings with sculptured heads. There is an upper cast-iron balcony at the side of the house. It is not known exactly when cast iron ceased to be used. It declined after the 1890s depression but was still being advertised in Lassetter's *Commercial Review*, no. 26 in 1911. Various patterns of 'Cast Iron Balcony Railings' were offered 'Height 2 ft. 9½ in. Weight, about 16 lbs. per ft. at 2/6 per foot.' (851 mm.-7.3 kg.) It was still in use in Lithgow, New South Wales, in 1919.

When the middle classes left the city for the new garden suburbs, the inner city developed a sleazy reputation. By the 1950s cast-iron was being pulled off terraces and replaced with fibro to enclose verandas to make an extra bedroom. In 1948 the Cumberland County Planning Scheme considered suburbs such as Woolloomooloo, Paddington, Balmain, Glebe, and Annandale to be obsolete areas and recommended the demolition and replacement of their housing stock within twenty-five years.

Return up Trafalgar Street and Collins Street to Johnston Street

Admire the spire of the Hunter Baillie Church from this perspective.

Edwinville, Trafalgar Street, Annandale

The facade of the Uniting Church, Johnston Street, Annandale

Continue right in Johnston Street

Note the small cottages opposite the Hunter Baillie Church. Despite modernisation, their lines can still be discerned. No. 59 has old wooden veranda posts, a door with leadlight glass, and columns at the windows.

By 1895 Annandale was known as 'a working man's suburb'. It knew the street cries of the 'prop man', the 'rabbitos'. and itinerant hawkers. Children played their games in the streets — chasings, hopscotch, rounders, statues, and skipping. Many of the younger people left the area for newer suburbs and in the years after World War II Annandale, like other inner Sydney suburbs, took on a derelict appearance. By the 1960s people began to realise the benefits of restoration. The Annandale Association was formed in November 1969, and there was a new awareness of the area, paralleling developments in Paddington, Balmain, and Glebe.

Continue along Johnston Street

Note terraces nos 67, 69, 71, and 73. Osborne has decorated tiles and plaster faces.

Council Chambers, 79 Johnston Street

Annandale is part of Leichhardt Council but it had its own council, Annandale Borough Council, when it seceded from the Leichhardt Borough in 1894. The Council Chambers date from 1899. A simple building, the entrance is guarded by two sculptured lions. Inside the main hall a wall plaque lists the mayors of Annandale, including John Young, Mayor from 1894 to 1896. Young purchased half the lots offered by Captain Robert Johnston in the North Annandale subdivision in 1876 (*see* p. 59).

Walk to the Uniting Church at 81 Johnston Street

The facade of the Uniting Church is a credit to the stonemason's art. The Methodist Church came to Annandale in 1883 and built a chapel in Johnston Lane before constructing the new church in 1891. The facade is of ornate Victorian stonework and the carved birds feed on stone fruit, and squirrels scamper among stone foliage. The church is brick. The stone facade was built in the 1870s in Sydney's Pitt Street. It was then part of Bull & Company's warehouse, which was gutted by fire in 1890. The windows have old geometric glass, now broken in places, and there are twin entrance porches. The church these days is a Uniting Church, and at the rear is a Memorial Hall erected during World War I.

Complete the walk up Johnston Street to Booth Street

Note the double-storeyed house at no. 83, the cottages at no. 85, and the residence at no. 87. All are typical of the Johnston Street architecture. Nos 91 and 93 have been restored and their old tiles and iron give pleasure to another generation. Next door the old stone and railing fence remain although the house has vanished and been replaced with brick apartments. Annandale is fortunate not to have lost too much of its character and past.

Cross Johnston Street to St Aidan's Church of England

St Aidan's, a modest brick church, was built in 1892. It has attractive stained-glass windows and now stands in the heart of the Annandale shopping area.

St Aidan was an Irish saint and the first bishop of Lindisfarne, off the coast of Northumberland. He had been a monk at Iona when King St Oswald of Northumbria made him bishop of the newly converted Northumbrians. The Anglo-Saxon theologian, the Venerable Bede, praised Aidan for his learning, charity, and simplicity of life.

Opposite St Aidan's Church is the Colonnade, which dates from 1894. This building was constructed by Adam Lappan, owner of a saddle and harness factory in Annandale. The original broad veranda and balcony of the 1899 Colonnade are missing. Annandale's first council meetings were held in Lappan's building.

Across Booth Street the old shops carry the date 1888, and Noonan's corner dates from 1909. In April 1987 an inappropriate metal rail construction was erected in front of these shops. The residents said it resembles a cattle truck. Ignore this incongruous 20th-century addition and take the opportunity to have coffee at a nearby cafe and you may feel like exploring more of Annandale. It is worth the effort to walk the length of Johnston Street to Rozelle Bay.

Annandale was a mixture of all classes of people. The comfortable middle classes occupied homes in Johnston Street, and artisans and workers had their homes near Johnston's Creek. It must have been pretty in the early days with the bush running down to Rozelle Bay and the creek meandering through the trees.

The 1870s subdivision was soon subdivided again into smaller lots and the cheaper land was acquired by the working classes.

The Annandale Theatre opened in Johnston Street near Booth Street in 1912. It was in the open air with timber seats; shows were interrupted if it rained. A new theatre was built around the old iron and timber frame and survived as the Annandale Picture Show until the Royal Theatre opened in 1928.

Continue down Johnston Street

It is easy to see why Johnston Street was thought to be the finest street in the colony. Its width and fine old homes still attract attention. Note nos 128, 132, 134, 129, and 127. Nos 147 and 149 have a fern-patterned cast-iron dating from 1885 and manufactured by Swinnerton & Frew. Cast-iron offered many motifs — there were Irish harps, shamrocks, true lovers' knots, parrots, birds, cupids, urns and flowers, and Flora, the Roman goddess of springtime. Note the decorative tiles at nos 155, 157, and 161. Those at no. 161 are a pretty daffodil pattern. No. 163 has old wide weatherboards and the little houses at nos 165 to 169, and 171 to 177 have high steps. Bay windows are a feature of nos 183 to 189.

A War Memorial dominates Hinsby Reserve. It is named for a former town clerk and was a park planned by the original Annandale Company. When the land was re-subdivided other park areas were cut up for housing. This park forms a pleasant square and there are fine homes fronting the area with a corner shop at the junction of View Street.

Proceed to North Annandale School

North Annandale Public School is built of brick in a pattern called 'blood and bandages' — alternate rows of sandstone and brick — and dates from 1905. It was a different world in 1905. The sexes were segregated at school and boys never played with the girls. Marbles dominated the playground — there were 'Aggies, Connies, and Clayies'. At recess boys engaged in games of big ring, little ring, or threes. Jacks were popular with girls. The knuckle bones came from the family butcher and 'mum' boiled them clean. Any boy worth his salt owned a shanghai (catapult) sticking out of his pants' pocket when not being used. Billy-carts were made from old fruit cases readily available at the greengrocers. In the classrooms the multiplication tables were chanted and the names of the kings of England learnt by rote. England was still 'Mother' to the Empire.

Continue past the school

Notice the different architectural styles. An old wooden shop still stands at no. 207 and some of the houses are being restored.

Winkworth Steps lead down to Rose Street

There is a view to the city and harbour bridge. As you walk on, the view on the left is to Glebe and the square tower of Toxteth House, now St Scholastica's Convent and school, is clearly visible. The view down on Rose Street reveals the old weatherboard backs to the cottages.

The Witches' Houses are a landmark of Annandale and were built between 1886 and 1889 by John Young.

John Young was an architect and engineer who settled in Melbourne in 1855. He found contracting more profitable than his profession and began to work in that role. He moved to Sydney in 1870. Young had been superintendent and draftsman of the Crystal Palace in London when aged only twenty-four and he worked on Sydney's Garden Palace, St Mary's Cathedral, St John's College, University of Sydney, and Sydney's Lands Department.

He became an important figure in Annandale's development. Near the waters of Rozelle Bay John Young acquired a nine-roomed Georgian cottage. It is believed the cottage was built by Captain Robert Johnston. Young extended the cottage to a comfortable villa, naming it Kentville for his home county in England. He established a bowling green at his villa and formed a club named the Sydney Bowling Club with the New South Wales governor as patron. Young was given the title 'Father of Bowls' in New South Wales. The club became the centre of the game in Sydney. Young did not confine his interests to bowls as Kentville also had a billiard room, quoit grounds, two shuttle alleys, and an archery field. Kentville was also noted for one of the finest gardens in Sydney.

Young died in 1907 and his estate was then divided into ninety small lots. Kentville was demolished. The estate is remembered in the naming of Kentville Avenue, a street of red-brick cottages.

It is believed the architect of John Young's 'Witches' Houses' was John Richards. The style is similar to homes built in the United States of America for wealthy citizens of the Victorian era but are unique in Sydney. The tall towers of the houses, which in the distance resemble witches' hats, gave the buildings their fanciful name.

The two northern houses, named Hockingdon, no. 264 and Highroyd, no. 262, were built for Young's daughters, Annie Reynolds and Nellie Daly, although they never lived there. Only one of the southern pair survives. This is Kenilworth, no. 260, and was the home of Sir Henry Parkes. Parkes died here in 1896.

No. 258 was Claremont. It was demolished in 1968. Its destruction led to the formation of the Annandale Association, and the house is featured in the society's emblem. Originally all the houses had balconies and verandas decorated with cast-iron. These have disappeared and the area has been filled in with unsightly additions. The houses desperately need restoration — there is even a tree suckling protruding from a tower.

Oybin next door at no. 270 is another former grand house. Architect Charles Blackman lived here between 1881 and 1885 and is generally credited with its design. John Sulman (later Sir John) bought a partnership in the practice of Charles Blackman. Blackman subsequently

*The 'Witches' Houses'
Johnston Street,
Annandale*

*Oybin, c.1880, once one
of Annandale's grand
houses*

The Abbey, an outstanding example of Gothic Revival style

absconded from the business and fled to San Francisco, USA, with most of the firm's money and a local barmaid. Sulman had found a note from Blackman stating he had left for New Zealand on urgent business. Sulman was later to become P. N. Russell Lecturer in architecture at the University of Sydney. He was also an opponent of a harbour bridge which he claimed could not 'fail to be a very prominent eyesore'.

On the corner of Johnston Street and Weynton Lane stands The Abbey. Most of the house is obscured by trees but note the fine stone fence and wooden Gothic gate. This house was also built by John Young, some say as an inducement to his wife to return to Sydney. Mrs Young had earlier gone to England. The Gothic Revival house was built between 1881 and 1883. The Victorian Gothic style had touches of medieval architecture with the pointed arch, oriel windows, decorative trefoils or quartrefoils, fretted timber barge boards and sometimes castellated parapets. The interior is lavishly tiled and stencilled. The Abbey is claimed to be the finest domestic example of the Gothic Revival style in Australia.

John Young was working at St Mary's Cathedral while he was building The Abbey and it is possible the house contains some of the stone from the earlier St Mary's destroyed by fire in 1865. There is a lion on the north gable holding a flag carrying the builder's initials, J.Y.

Gateway to the Abbey

Stroll down Johnston Street to view Rozelle Bay

There is a fine panoramic scene of the city and the harbour bridge. You will also notice the old railway line and the brick aqueduct. Across the bay the red-brick St Augustine's Church at Balmain is clearly visible (see p. 37).

The citizens of Annandale and Glebe rallied to preserve the foreshores of Rozelle Bay for an area of public recreation. Federal Park is on the Annandale side of Johnston's Creek. The estuary and foreshore area had been reclaimed by the Department of Works in 1898 and 1899. At that time Annandale had fought to prevent industrial development and the park was proclaimed in 1899. On the Glebe side of Johnston's Creek is the area is known as Jubilee Park. At the turn of the century it was a popular Sunday afternoon walk to Glebe Point. The foreshores were leased to industry, but the citizens of Annandale may win the fight to have the whole area made a recreational ground.

Return up Johnston Street to Winkworth Steps

Wander from Johnston Street down Winkworth Steps to Rose Street.

Continue along Rose Street and cross View Street

Note Ascot Terrace at the corner of Rose Street. It has been restored and reveals something of its original appearance. At the junction of Rose and Trafalgar streets there is an old corner shop. It has decorative urns, a cast-iron balcony, and bears the date 1906.

Study the architectural styles of the houses on the high side of Rose Street — nos 30, 28, and 26. No. 28 has an unusual door knocker. No. 26 is two-storeyed. Note the tower and iron railing balustrade. At no. 24 there is a pretty iron railing to the steps. Nos 16 and 14 are double-storeyed brick with iron work and old stone gate posts. Notice the barge boards on nos 12 and 10 Rose Street.

From this section of Rose Street one looks down on Harold Park Paceway (*see* p. 88) and the tower of Toxteth House in Glebe.

Turn left from Rose Street into Nelson Street

Johnston's Creek divides Annandale and Glebe. In the days before settlement it wound its way through native bush and tall gums. Today it is relegated to the role of a storm-water channel.

The Water Board Aqueduct dates from *c.*1896 and was designed by Cecil West Darby. Its job was to carry sewage to the Bondi Outfall. It is classified 'O' by the National Trust and is constructed of Monier reinforced concrete — the first use of this material in Australia.

Nelson and Trafalgar streets were named by Captain Robert Johnston who had served with the British Navy in the Napoleonic wars and he chose to remember the great victory of Trafalgar in his Annandale estate. Nelson Street has small semi-detached cottages and two-storeyed terraces. Note the old tiled shop at no. 315 and the tiny cottage at 313. On the right there is a high rock wall. The cottages at nos 307 and 301 are constructed of old wide weatherboards. Wander past the houses noting the different styles — 209 is a two-storeyed terrace; 289-287 are brick terraces; 285 is a cottage; 266 two-storeyed terrace with patterned tiles; and 262 is a brick semi with blue plumbago overhanging the fence.

Annandale Street is the most highly desired residential address in the suburb, but properties in Johnston, Nelson, and Trafalgar streets are also sought after. Nelson Street to Piper Street has old weatherboard cottages, with wooden fences and no. 260 an attic in the roof.

*Turn right from Nelson Street and climb steps to **Piper Street***

To the east are modern town houses for the new residents of Annandale.

*Turn left into **Trafalgar Street***

Again you will notice a street of semis, terraces and weatherboard cottages. Note barge boards at no. 208, nos 173 to 171 two-storeyed brick residences with towers and decorated urns and the A-line barge boards at no. 169. Near the new brick town houses in Trafalgar Street there is a view to the University of Sydney. It is interesting to study the mixture of housing in Annandale. Trafalgar Street lacks the fine houses of Johnston Street. The land in this area towards the creek was cheaper and so more of the working people dwelt here.

*Turn right from Trafalgar Street into **Booth Street***

You may end your walk by browsing in Booth Street from where there are buses back to the city.

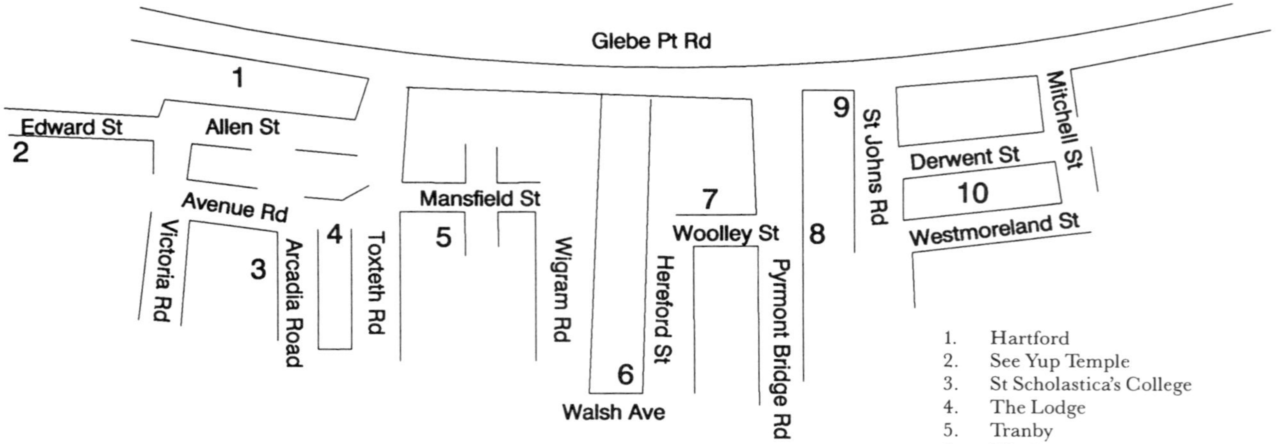
Glebe Pt Rd
1
Edward St
2
Allen St
Avenue Rd
Victoria Rd
3
Arcadia Road
4
Toxteth Rd
Mansfield St
5
Wigram Rd
6
Walsh Ave
Hereford St
7
Woolley St
Pyrmont Bridge Rd
8
9
St Johns Rd
Derwent St
10
Mitchell St
Westmoreland St

1. Hartford
2. See Yup Temple
3. St Scholastica's College
4. The Lodge
5. Tranby
6. Kirribee
7. St James's Church
8. The Abbey
9. St John's Church
10. The Glebe Estate

Walk No. 5

Glebe

The name 'glebe' comes from ecclesiastical law and refers to 'the land devoted to the maintenance of the incumbent of a church'. It comes to the English language via the French and from the Latin *gleba (glaeba)* meaning 'clod, turn, land or soil'. It was first used in England in the 14th-century.

In 1789 Governor Arthur Phillip received 'Additional Instructions Regulating the Allotment of Land for Church and School Purposes'. In 1790 Phillip ordered a survey of 1,000 acres (405 ha.) at Petersham Hill, also known as the Kangaroo Grounds. The surveyed land extended from Blackwattle Bay to the present site of Darlington. To the north 400 acres (162 ha.) was declared 'Church and Glebe Land', a central section of 400 acres became Crown land, and the southern area of 200 acres (81 ha.) was set aside to support a schoolmaster, although in 1789 there was no schoolmaster in the colony.

The Reverend Richard Johnson, first chaplain of the colony, arrived with the First Fleet, and held the first church service on Sunday 3 February 1788 in the shade of 'a great tree'. He built the colony's first church; its site is now marked by a sandstone memorial at the junction of Bligh and Hunter streets, Sydney. The Reverend Johnston did not regard the infant settlement of Sydney or its inhabitants very highly. In November 1788 he wrote that the Government

> would act very wisely to send out another fleet and take us all back to England, or to some other place more likely to answer than this poor wretched country, where scarcely anything is to be seen but Rocks, or eaten but Rats.

Nevertheless Johnston was soon farming and acquiring stock. He planted orange pips acquired at Rio de Janeiro and corn and vegetables

Glebe Point Road from St John's Road, c.1882. The scene is still recognisable today, although the church spire on the horizon was a figment of the artist's imagination (ML)

and subsequently remarked on his flourishing 'little garden' in Bridge Street, Sydney.

Governor Phillip gave Johnson the 400 acres (162 ha.) of land known as The Glebe or St Phillip's Glebe and assigned him three convicts to clear it in 1789. The Reverend retorted, 'Four hundred acres for which I would not give four hundred pence' and requested the usual 100 acres (40.5 ha.) grant in its place. He named this grant Canterbury Vale. (It is the site of the present suburb of Canterbury.) The grant was later increased to 350 acres (142 ha.). Johnson also owned two acres (0.8 ha.) on 'the Brickfield' (now near Central Railway) and 120 acres (48.5 ha.) at Ryde, which he sold in March 1800.

The glebe land considered unsuitable for agriculture by the Reverend Johnson remained virtually untouched for the next thirty years. It was heavily forested and had 'the brightest parrots'. The timber was cut on occasions and the rocks and bays frequented by fishermen. The water from the Blackwattle Swamps was carted to Sydney town to augment its water supply.

By the 1820s free settlers were arriving in the colony in increasing numbers. There was a need for more roads, schools and churches, and the Church of England founded a corporation — the Trustees of Clergy and School Lands in the Colony of New South Wales. The corporation was to acquire and dispose of land, and funds were to be used to build the roads, churches, and schools.

In 1828 the Glebe was subdivided into twenty-eight allotments and offered for public auction. Lots 7 and 8 were given to the Trustees of St Philip's, Church Hill, to raise income for the diocese, and Lot 28 was to provide income and a residence for Bishop Broughton. It became Bishopthorpe. Following the subdivisions, the merchants and wealthy began to build substantial homes at Glebe.

Our walk commences at the corner of Toxteth and Glebe Point roads

The streets of Glebe are lined with mellowed Victorian cottages bordered by iron railings or wooden fences. There are decorative tiles, etched glass panels, and fanciful chimney pots. Wander along Glebe Point Road noting particularly nos 234 and 244. These are fine old homes, still cared for and loved. No. 234 has been fully restored and the garden planted. No. 244 is Hartford — it dates from 1899. It has a lead-roofed turret of the Federation style.

Turn left down laneway at side of no. 244 to Allen Street

Note no. 13 on the corner with its pretty lace veranda and carved barge boards.

Turn right down Allen Street

Look at nos 34 and 32 with their towers, barge boards, slate roofs, and coloured glass. As you wander along the street note the veranda tiles and decorative wall panels. Can you find the lion statues and plaster garden boys?

Turn left at Victoria Road and right into Edward Street which leads to the Chinese joss house

Sze Yup Temple. It is surprising in the middle of Victorian Glebe to discover a Chinese joss house. Like other areas around Sydney, Glebe at one stage had its Chinese market-gardens. The Chinese had come to Australia at the cessation of convict transportation to New South Wales in 1840. They were seen as a source of cheap labour, which was then in short supply.

After the goldrush period some settled and engaged in market-gardening. By 1900 half the market-gardeners in Australia were of Chinese origin and the Chinese were a familiar sight in city suburbs before World War II. Their produce was sometimes carried by a horse and cart but more usually in two baskets suspended from a pole across the shoulders. Many of the Chinese came from the Canton Delta of Kwangtung, and others from the Zse Yup region of Canton. The latter acquired apartments buildings and stores in Dixon Street, Sydney, and built a joss house in Glebe near their market-gardens.

The land for the joss house was purchased from the wife of a local hotel owner, Mrs Elizabeth Downes, in 1897 for £325 and the joss house was built the following year. The name 'joss' comes from a corruption of the Latin *deus* meaning 'god'. The original joss house was burnt down and a new one, the Sze Yup Temple, opened in March 1955. There was a traditional celebration by the Chinese community with firecrackers and splendid Chinese papier-mache dragons. The impressive red and green tiled gateway with ceremonial lions was opened by the Mayor of Leichhardt, Alderman Evan Jones, on 15 May 1893.

Walk through the gateway of the green tile topped wall into the grounds and note the Keelins, benevolent mythological figures, on the roof of the temple.

The temple consists of three buildings: the central one is dedicated to Kwan Ti, god of war, A military leader born in 221 AD, he died in 265 AD fighting for the emperor. Kwan Ti became a god in 1594 AD and is still the patron of Chinese soldiers. The remaining buildings honour gods of fortune and the dead. To the left is a shrine to the early Chinese settlers in Australia.

Above, *the gateway to the Chinese joss house, below*

Worshippers visit the temple for Chinese New Year and other traditional ceremonies when incense is burnt and offerings of food made; imitation paper money is also burnt in iron stoves.

Return up Edward Street and turn right at Victoria Street and left at Avenue Road

As you walk up Avenue Road note the different style of cottages and look for the cast-iron friezes and wooden barge boards. No. 4 is Federation style and is named Wychwood — look at its decorative plaque featuring flannel flowers and waratah. Opposite no. 25 is a pretty house with an attractive lacy iron veranda.

Continue to corner and St Scholastica's College

After the first subdivision of land in Glebe, the larger landholders employed architects to design imposing houses: Lyndhurst, Hereford, Boissier, Eglintoun, Forest Lodge, and Toxteth. The main building of St Scholastica's College was originally Toxteth House and was designed by architect John Verge for George Allen (*see* Profile).

George Allen arrived in Australia in 1816 from London when he was only sixteen. His father was Dr Richard Allen, a physician to the Prince Regent. Dr Allen died in 1806 and his widow married a business colleague, Thomas Collicott. He was convicted in 1812 on a charge of failing to affix revenue stamps to medicine bottles and transported to New South Wales. A family friend, Sir Robert Wigram, booked passages for Collicott's wife and family to New South Wales and provided a letter of introduction to Governor Macquarie. Macquarie arranged for George Allen to be articled to the Crown solicitor. He became the first solicitor to train in Australia and was admitted to practice in 1822. Allen owned considerable land at Blacktown which he sold to purchase land in Surry Hills. In 1823 he married Jane Bowden, the sixteen-year-old daughter of the schoolmaster of the Male Orphan School. Their son, George Wigram Allen, was born at Surry Hills in 1824. That year Allen sold his Surry Hills property. He later owned three Sydney houses as well as property at Botany Bay and his Glebe estate.

John Verge designed a white stone Regency villa of two storeys with a single-storeyed veranda and fluted pillars. The house was built between 1829 and 1831 and named Toxteth Park after the English home of the family's benefactor, Sir Robert Wigram.

Long French windows opened onto the flagged veranda and the grounds were extensive. There were surrounding paddocks and orchards. Harold Park Racecourse now covers part of the orchard. There were fifteen convict servants, a free overseer, seven horses, and 150 head

St Scholastica's College, Glebe formerly Toxteth House

of cattle. George Allen had a full and busy life and was a leading member of society. He built a Wesleyan chapel in the grounds of Toxteth Park. When he died in 1877 his son, Sir George Wigram Allen, added a third storey to the central block of the house, a tower, and Italianate embellishments. The work was carried out between 1877 and 1881 by architect George Mansfield.

Sir George Wigram Allen was also a solicitor, member of the Legislative Assembly and Legislative Council, and Commissioner of National Education from 1853 to 1866. He was first mayor of the Municipality of Glebe and was elected for 18 consecutive terms. He was knighted in 1884, one year before he died at Toxteth House in 1885.

In 1904 Toxteth House was purchased by the Roman Catholic Church, and the Sisters of the Convent of the Good Samaritan came to Toxteth. Their old convent in Pitt Street was demolished to permit the construction of Central Railway.

It is permissible to peep through the stone gateway of St Scholastica's to admire Toxteth House. Note the tower of the house. Flags were once hoisted here to announce ship arrivals in Sydney Harbour. Toxteth House could see the house flags of ships flown at the Sydney Observatory.

The sun dial in the garden has a spelling error in its inscription:

> Presented to Geo. Allen Esq. of *Toxeth* Park MLC by J.D. Callaghan, Mathematician for Lat. 33O 56' Sth A.D. 1857.

There was a story that a ghost haunted the grounds of Toxteth House, but it was discovered that the 'ghost' was a white peacock taken to night-time wandering.

Turn right and wander down Arcadia Road

The sun warms the streets of old Glebe and there is a profusion of trees in the street and gardens. Arcadia Road is lined with charming Victorian cottages. There are curlicued barge boards, glistening floral wall tiles, plaster festoons and urns, quaint chimney pots, and sturdy black iron fences. The mood encourages one to stroll leisurely and admire a rose, a bush of mauve lavender, or a cat snoozing on a sunny porch. Note the different architectural features of the houses as you wander to the end of Arcadia Road.

At the road's end there is a view across to Annandale — look for the spire of the Hunter Baillie Church and the famous Witches' Houses (*see* Annandale Walk).

Retrace your steps to Maxwell Road turn right and walk to Toxteth Road

The houses in this area stand on the original land of Toxteth Park and it is hard to visualise the sheep and orchards that once existed here. Toxteth Road overlooks Harold Park Trotting and Greyhound Racing Track. It was named for Childe Harold — not the hero of Byron's poem, but a famous American stallion and sire that set many trotting records in New South Wales. The track was earlier known as Forest Lodge Track and was home to the New South Wales Trotting Club which was founded in 1902. The club moved to Kensington racecourse after its first two meetings but returned to Glebe in 1904. The track then became known as the Epping Racecourse. The racetrack was originally leased from the Metropolitan Rugby Union but was purchased by the Trotting Club in 1911. The greyhound track was built in 1927 and in 1929 the named changed to Harold Park because Epping Racecourse was being confused with the northern Sydney suburb.

One wonders what George Allen would make of his estate these days. A man of strong Methodist and anti-gambling principles, his old home is now a Roman Catholic College and his orchard covered by a race track.

On the corner of Maxwell Road and Toxteth Road note the Federation house — no. 43 Toxteth Road — and the veranda tiles and tiled wall panels featuring a waterbird among bullrushes and an urn of flowers.

Turn left up Toxteth Road

Look for features such as iron lace and plaster work. Note no. 39, terraces at nos 46-44, and the large two-storeyed residence at no. 40. It has iron columns, decorative lace, barge boards, and a castellated pediment. Note also nos 29, 25 and 23.

Pass the Uniting Church. The Hutchinson Memorial School Hall is dated 1898 and boasts elaborate stone carving over the doorway. Pass Oruba, a cottage at no. 17, and look also at no. 15 with its slate roof, iron columns, and frieze, and no. 11, Lucia.

Proceed to the junction of The Avenue and Mansfield Street

At the corner of The Avenue is The Lodge, the gatekeeper's cottage to the Toxteth Park Estate. It was designed by George Allen Mansfield and built about 1877 for George Wigram Allen when he was busy adding to Toxteth House. It is a charming Gothic Revival stone cottage — look at the porch of stone and carved wood.

The colony was self-governing in the 1870s. The British Army had vacated the colonies which now had to raise their own defence forces. It was a time of considerable expansion, and architects Edmund Blacket, James Barnet, Thomas Rowe, and John Young were designing imposing buildings in Sydney: the Lands Department, the Great Synagogue, St Andrew's College at the University of Sydney, and Blacket's lovely St Stephen's Church at Newtown.

In 1879 the great international exhibition was staged at the Garden Palace, designed by Barnet. T. S. Mort's company successfully exported the first frozen meat to London in the same year, and many commercial buildings were built in Sydney. Among them was the City Bank in Pitt Street with 'a gorgeous spiky roof of turrets and finials done in the French manner', which was built by two architect brothers, the Mansfields. The name is commemorated in Glebe's Mansfield Street.

Cross to Mansfield Street

George Allen Mansfield (*see* Profile) at one time lived in The Lodge of Toxteth House. He was architect to the Council of Education and designed many public schools between 1867 and 1880. He also designed the Royal Prince Alfred Hospital and was a founder of the Royal Australian Institute of Architects. He was its president for the first three years. The Mansfields and Allens were connected by marriage: George Allen Mansfield married Emma, daughter of George Wigram Allen.

At the corner of Mansfield Street is no. 27, Endersley — a gracious old home. It is asymmetrical — not having corresponding parts. Note wooden veranda posts and frieze, the shutters, and fine stone and iron fence. Many of the Glebe houses have been painted in heritage colours — no. 25 has been restored beautifully and is a warm orange ochre with plum iron work. Note the lovely etched glass.

By the late 1830s there were cottages and small factories along Glebe Point Road and developments had begun to encroach on the great

The Lodge, formerly the gatekeeper's cottage of Toxteth House

estates. By the 1840s more cottages were built and people were leaving the crowded conditions of the city. The boom occurred between the 1860s and the 1890s when dwellings in Glebe increased in number from 720 in 1861 to 3,225 by 1891. By the outbreak of World War I Glebe was fully built out.

Cross Boyce Street and note Tranby — no. 13 Mansfield Street

Tranby may date back to the 1850s and is of a post-Regency design. Note the flat iron columns on the veranda.

Tranby was once home to both the Boyce and Mansfield families. George Wigram Allen had married Marian Boyce, daughter of the Reverend W. B. Boyce, a member of the first Senate of the University of Sydney. There is an old photograph of Tranby surrounded by open ground, low shrubs, and tall trees. It has had a varied history. At one time it was a hostel for the University of Sydney and home to the Reverend John Hope of Christ Church, St Laurence. The Reverend Hope presented Tranby to the Aborigines Co-operative of the Australian Board of Missions in 1958. Students are trained for various careers and as promoters of Aboriginal co-operatives.

Look at the restored house opposite Tranby on the corner of Boyce Street. It has dark green iron and an orange tree grows in the garden. Nos 18 and 16 have tiny cast-iron balconies. A creamy briar rose tumbles over a dividing iron fence.

Note the houses, nos 1-7, to the corner of Mansfield Street. These have iron posts and a wooden lattice frieze.

Proceed to Wigram Road

In Wigram Road, named for George Wigram Allen, note Minerva Terrace — nos 11A-17. It is a late Victorian terrace of the 1890s with fine plasterwork. Note the chimneys and parapet. The Victorian Italianate style often has a panelled door with sidelights and etched glass and triple light arched windows with hood mould and decorative plaques. Parapet walls have Italianate balusters, urns or, perhaps, Rococo shell decoration.

Many terraces were built in the 1870s and followed the style of London, often being on narrow blocks with rear lanes.

Turn right in Wigram Road

Wigram Road offers many architectural styles. Some houses appear to be neglected or to have been 'converted' in an inappropriate manner by, for example, enclosing balconies or incorporating aluminium window frames. Note the terraces on the right, nos 28-40.

Note the terraces opposite Walsh Avenue. They have pretty cast-iron and charming tiles. Note the cottage at no. 48 — it has pretty tiled steps, and an old lemon tree is growing in the grounds.

Turn left into Walsh Avenue

There are 1950s brick flats on the right.

Turn left at Hereford Street

Hereford Street runs down to the Royal Alexandra Hospital for Children. Few would realise there is a creek at the bottom of the street, Orphan School Creek, no doubt a sylvan stream in earlier days.

Turn left in Hereford Street

At no. 55 is Kirribee, designed by James Kirkpatrick in 1889 for Joseph Leeds. Kirkpatrick succeeded Thomas Rowe as architect on Rowe's Sydney Hospital in Macquarie Street. Kirribee is a typically elaborate Italianate house and is said to have the best stained-glass windows in a domestic hose in Sydney. Look at the grand entrance, iron columns and railing. It is now occupied by the New South Wales College of Nursing. Next door is Hereford House, which was built in 1874-75 by William Bull and is now the Nursing Education Centre.

This is not the original Hereford House built by architect Edward Hallen for George Williams in 1829 and described as an 'elegant house'. That building stood on an elevated site covered today by the park at the junction of Glebe Point Road and Bridge Road (*see* p. 94). In 1851 that Hereford House was owned by Thomas Woolley, a wealthy ironmonger. With his death in 1858 the estate was subdivided and sold.

Tennis enthusiasts will be interested to know famous Glebe tennis star Lew Hoad once played on the courts at the rear of Hereford House.

Continue up Hereford Street

Times were often hard in Glebe. Boys left school as young as twelve years of age to seek employment and support their families. Girls had to work in local shops or factories, perhaps at Grace Bros. on Broadway. The boys' opportunities were greater: there were coal and timberyards, abattoirs at Blackwattle Bay, wheat-loading at Glebe Island, bakeries, the White Bay powerhouse, or the tram depot at Rozelle.

It was harder in the Great Depression when most of Glebe's inhabitants were poor. Shelters were organised by Canon R. B. S. Hammond of

St Barnabas' Church on Broadway. In 1938 he bought a house named Rosebank in Hereford Street as a shelter for needy families. The houses were known as 'Hammond's Hotels' — his first shelter had been in a former old Glebe home, Tressmanning.

There were good times too, and author Frank Clune in *Saga of Sydney* recalled wagging school to go to Blackwattle Bay to 'catch "yellow-bellies" [a small fish] on its muddy shores as the tide ebbed and flowed'.

Note the house on the left — it has black and white veranda tiles and green lace columns and railings. The house next door is painted pink with white iron lace, and Kinkara Mews, nos 43-47, is built in a sympathetic style.

Turn right into Woolley Street

The street is named for ironmonger Thomas Woolley. You will pass St James Primary School where feelings ran high at the time of World War I. One class was told by an Irish priest it would be a good thing if England lost the war as Ireland would fare better under German rule. One Glebe lad announced this sentiment at dinner that evening. His father was English, his mother of Irish-descent, and his older brothers were fighting in Gallipoli and France. His father was outraged by this statement and the lad and his younger sister were promptly removed from the Catholic school to attend the local public school. No amount of entreaty by the priest and nuns to the lad's mother could revoke the father's decision. Today another generation of children play in the grounds.

Pass landscaped St James Park and cross to St James Presbytery and St James Church

The Presbytery has black iron lace and a stone and spear iron fence. The church opened in 1877. Previously Catholics from Glebe had to attend services at Balmain. The church of plum-coloured brick was designed by James Kirkpatrick. Inside note the fine timber ceiling and the southern rose window. Unfortunately the church is marred by an ugly metal porch. From the steps of St James one looks across to the spire of the old Presbyterian Church which is now a restaurant.

Walk to Bridge Road

Note the house on the left of Bridge Road. There is a picket fence with turned wooden posts.

Cross to The Abbey

Architect Thomas Rowe designed a Presbyterian church in High Victorian Gothic style for a site on Parramatta Road. The church was built between 1876 and 1881, but the noise of the traffic (especially the trams) was considered too great for congregations to tolerate, and in 1927 the church was re-erected at 158 Bridge Road, where Alma Cottage stood. The house was demolished to make room for the church. Another house, Hamilton, was incorporated in the church hall, and Reussdale, designed by Ferdinand Reuss Snr, became the church manse. In 1977 the church became The Abbey Restaurant.

Reussdale still stands in the grounds of The Abbey, a lovely old house, totally neglected, and fast falling into a state of ruinous decay. Reuss designed the house in the style of the 1880s; it has gabled ends, timber barge boards, and a fine bay window.

Opposite at nos 181-179 are twin houses lovingly cared for. In the garden are a white magnolia, tree ferns, a palm and native gums. These houses were also built by Reuss.

Walk up Bridge Road towards Glebe Point Road

Next to The Abbey is an Italianate residence, The Hermitage, hidden behind a high wall. Designed by Ferdinand Reuss Snr, it was at one time his home; his son also lived there until 1925. The terraces at nos 152 and 150 have a true lovers' knot pattern of iron work.

Turn right into the park

The park at the junction of Bridge Road and Glebe Point Road covers the site and grounds of Hereford House. It was completed in 1829 by Edward Hallen, architect of the Sydney College, now Sydney Grammar School. John Verge designed the servants' quarters. It was built for Attorney-General John Kinchela and had views of Darling Harbour and the Macquarie Lighthouse on South Head.

Kinchela resided at Glebe for only one year and then rented Juniper Hall in Paddington. A chemist, Ambrose Foss, then acquired Hereford House for £2,000 but later built his own house, Forest Lodge, nearby (now occupied by 208-210 Bridge Road). A merchant named Hirst purchased Hereford House but he suffered financial losses in the 1840s depression. The house changed hands many times: Thomas Breillat of Huguenot descent, owned it in the 1840s, and Thomas Woolley lived there in 1851. After Woolley's death in 1858 the estate was subdivided and the house became the home of Judge Wilkinson. He used to drive by carriage each Sunday morning to St John's Church, immediately next door to Hereford House. The judge died at the house in 1908, and in 1910 it became a teacher-training college and later a residential college. Hereford House was demolished in the 1920s. An old photograph shows the colonial mansion of two storeys with an elegant veranda. There were tall trees in the grounds. It is said the dining and drawing rooms were each thirty feet (9 m.) long and there were French windows, marble fireplaces, and 'handsome stoves'.

Sydney has lost many gracious homes but Lyndhurst, another Glebe colonial home, built for Dr James Bowman, surgeon of the Rum Hospital, has been partially preserved in Darghan Street where it is the home of the Conservation Resource Centre of the Historic Houses Trust.

As you leave the park note the elaborate War Memorial erected in 1921 in Glebe Point Road.

Climb the steps to the Church of St John the Evangelist, Bishopthorpe

Glebe consists of two areas dating from the 1828 subdivision — St Phillip's and Bishopthorpe. The Church of England parish formed in 1856, and the first church opened in 1857. A temporary church had been built on the east side of St John's Road and was used until 1870. Queen Victoria's son, the Duke of Edinburgh, was to set the foundation stone of St John the Evangelist. But an attempt on his life at Clontarf, Sydney, on 12 March 1868 prevented him from doing so, and the stone was laid by the Governor of New South Wales, the Earl of Belmore, on 15 April 1868.

The church is the work of architects Edmund Blacket and Horbury Hunt. At the time Hunt was employed in Blacket's office. Of Victorian Romanesque style and built of stone from Pyrmont quarries, the church was completed in 1870. Horbury Hunt wished to erect iron columns inside but was over-ruled by Blacket who favoureded rows of fine arched stone columns that now grace the interior. The church contains some lovely wooden work; the plain stained glass is designed to focus attention on the altar. The tower of St John's, the Tovey Memorial Tower, was designed in 1911 by Blacket's son, Cyril. It is 100 feet (30.5 m.) high and holds thirteen bells which were cast in Sydney, the first peal to be made in Australia, by W. Taylor's foundry at Pyrmont. They were first rung on Christmas Eve, 1911. The old 1873 rectory, which stood nearby, was demolished in 1963. The site is now covered by St John's Church of England retirement village. The village was designed in 1964 by architects Hely, Bell & Horne and won the Royal Australian Institute of Architects Sulman Medal.

As you leave the precinct of St John's note the fine stone wall with its lychgates. It was built in 1901 as a result of the efforts of the then Rector, Reverend S. S. Tovey.

Cross St John's Road to the Parish Hall

Opposite St John's Church is the Parish Hall. It was designed in 1897 by Edward Halloran and commemorated Queen Victoria's Diamond Jubilee. It is known as Record Reign Hall. The coloured brick building is decorated with a terracotta plaque bearing the Queen's portrait. It carries the inscription:

> From my heart I thank my beloved people.
> God bless them.

There is another plaque inside the hall with the message:

> May the children of our children say she wrought her people lasting good.

Terracotta was popular at this period and the Sydney kilns were fully utilised. Architect George McRae employed the medium in 1892 by applying terracotta spandrils and shield to his Eastern Market at Woolloomooloo.

You may care to see more of Glebe by exploring a little of the Glebe Estate

The Glebe Estate comprised the lands held by the Church of England. The 1828 subdivision offered twenty-eight allotments for public auction but reserved three adjoining lots. Two of these — Lots 7 and 8 — were handed over to the Trustees of St Phillip's (later St Philip's), Church Hill, Sydney, and a third lot — Lot 28 — set aside for the residence and income of Archdeacon Broughton in 1836 when he became Bishop of Australia. This area was named Bishopthorpe and with the St Phillip's Estate covered nineteen hectares — the Glebe Estate — and mostly was owned by the Church of England. It was roughly triangular in shape and was bordered by St John's Road, Arundel Street, and Cowper Street to Wentworth Park Road. After World War II urban administrators planned to clear old inner suburbs. The County of Cumberland Plan was gazetted in 1951 and contained a proposal for the North-Western Expressway to pass through Glebe and Pyrmont. The Western Distributor was to cut across the Glebe Estate. In 1968, when local government boundaries changed, the Glebe Estate passed from the City of Sydney to the Municipality of Leichhardt.

The Glebe Society was formed in 1969 and by 1970 was agitating for controlled residential development. In 1971 the New South Wales State Planning Authority produced its Glebe Study and the same year the Church decided to sell the Glebe Estate.

Local Government elections in August 1971 brought about a change in the composition of Leichhardt Council with an independent reform

Note the terracotta plaque on Record Reign Hall

group backed by the Glebe Society being elected to the new 'Open Council'. In 1972 the Church of England offered its residential estates to the Government to enable the areas to remain as low income housing. At the end of 1973 the Commonwealth Government negotiated the acquisition of the estate.

In 1974 the Commonwealth Government under Prime Minister Gough Whitlam acquired the Glebe Estate for $17.5 million and the Glebe Project Office opened on 12 August 1974. An area of nineteen hectares containing some 730 houses were to undergo urban renewal. In 1985 the Commonwealth relinquished its role in the renovation of the Glebe Estate and its administration passed to the New South Wales Department of Housing.

Tradesmen and artisans made their homes on the estate and its architecture varies considerably from other areas of Glebe.

From Record Reign Hall walk up St John's Road to Westmoreland Street

Note the Fire Station built in 1906 and designed by architect Walter L. Vernon in St John's Road. There are pretty barge boards at nos 77-79 and a row of simple cottages from Westmoreland Lane to Westmoreland Street.

Turn left at Westmoreland Street

On the right-hand corner of Westmoreland Street is a two-storeyed Italianate house. The iron balcony has a pattern of a basket of lily leaves.

Chesterfield House, built in 1875

Note the plaster work and the palisade fence topped with arrow heads surrounding the small leafy garden.

The depression of the 1840s caused by a disastrous drought also resulted in some of the large land holdings in Glebe being cut up. In 1842 the area of St Phillip's comprising Lots 7 and 8 was divided into thirty-two one acre (0.4 ha.) allotments and workmen's cottages began to spot the landscape. Westmoreland Street is lined with leafy trees but in the 1860s small cottages sat scattered in empty paddocks at the outskirts of the city. In the early 1870s two-storeyed terraces with verandas and balconies were being erected. Nos 66 and 64 have pretty wooden barge boards and note the wooden veranda frieze at nos 60-56.

No. 61 Wynthorpe dates from the 1870s. No. 55 is two-storeyed with a balcony. In 1875 it was an old inn, the Toxteth Park Hotel. The present laneway was once a carriageway to the rear of the hotel.

No. 41 is a two-storeyed house set amid the cottages. Before World War I it was occupied by John Hackland and family. Born in London in 1865, John Hackland had gone to sea when a lad and risen to the rank of first mate. He was on the Australia run with the Duncan Moore line of ships when he met a girl of Irish-descent in Sydney. They married at St Mary's Cathedral in 1890 and John left the sea. He owned shops in various Sydney inner suburbs and later set up the Hackland Carrying Company in Glebe. The business was expanding when World War I interrupted normal activities. His sons enlisted and a daughter, Mary Machon, recalls the day her younger brother came home a soldier. She watched him walking up Westmoreland Street. All the neighbours came out to pat him on the back and shake his hand. Brother Jack was only sixteen but

Restored Glebe Estate cottages in Westmoreland Street

had lied about his age to go off and fight. He was to serve at Anzac Cove and in the trenches of France. At war's end he came home to Sydney but his lungs were scarred with mustard gas from the battlefields of France.

Note the pretty iron on no. 41. It is a circular pattern featuring a true lovers' knot at the centre. The design was registered in Victoria in 1872 by E. Cross and P. Laughton and the Sydney hardware store, F. Lassetter & Co. were still offering the design as late as 1914:

> Height 2 ft. 9½ in. Weight, about 16 lbs. per ft. 2/6 per foot. [85 mm.-7 kg.].

There is a fine hip-roofed 1881 cottage at no. 15. Nos 4-10 is the earliest terrace in the area with a balcony over the veranda, built in 1869.

No. 2 is Chesterfield House built in 1875 on a triangle of land. It is a mixture of post-Regency and Italianate styles. Note the steep moulded gables, narrow windows, and tall chimneys. During restoration of the area, many chimneys were found to be in poor repair, often cracked and with stucco falling off and some even structurally faulty.

Turn into Mitchell Street and walk towards Glebe Point Road

No. 136 is a glass factory. Glass-blowers are busy at work and samples of their craft are available for sale. Note the old corner shop at no. 46. Opposite at no. 65 Derwent Street is Demeter's Bakery. The bakery was built in the late 1870s but the upper storey was added in 1894. When the estate was acquired by the Commonwealth Government the old brick ovens were still intact. The bakery had also been a pie shop and hundreds of pie tins were found on top of the ovens.

Demeters is run by Helios Enterprises Ltd and is a registered non-profit organisation — a prototype based on the teachings of Austrian philosopher and teacher, Rudolf Steiner. The bakery has operated for more than fifteen years and staff, full- and part-time, have a say in business decisions, and wages are paid according to need.

On the first Saturday of each month open baking sessions are held. Bio-dynamic agriculture is discussed during the baking and while your own loaf of bread is baked a wholesome lunch is provided. All types of groups are welcome, and children particularly get involved in kneading the dough. The smell of baking bread is delicious.

Derwent Street is on land once owned by architect Edmund Blacket. Blacket lived in Glebe from 1853 to 1870 and was one of its many famous residents.

Note no. 52 — an old two-storeyed house. When the Commonwealth Government purchased the area, the veranda and balcony had been filled-in with windows and cooper-lourvres. The house was partly restored and old cast-iron and iron posts once more grace the house. There is a slate roof and plaster work over the arched windows.

From the corner nos 69-75 are a row of restored cottages — note the traditional striped iron over the verandas. No. 79 dates from 1869 to 1870; nos 83-81 were built in 1868. Note the picket fences to the houses in Derwent Street. The original fences throughout the Glebe Estate were in a bad state of repair and they have been restored in Victorian and Edwardian style. There are pretty house number plates featuring the Glebe Project logo.

Wander up Derwent Street noting the different styles and colour schemes. All the Glebe Estate houses have been painted, upon restoration, in heritage colours. There are shades of vanilla, cream, biscuit, stone, crimson, dark brown and deep Brunswick green. Also note the iron veranda canopies.

In the *Sydney Morning Herald* of 14 February 1850, E. C. Weeks & Co., Ironmongers, of 450 George Street advertised

> Galvanized Sheet Iron, For Roofing, Verandahs, &c. It is considerably cheaper than lead, and being incorrodible, is in general use for guttering, rainwater heads, pipes, &c. In England it is used as roofing to all Government buildings, railway stations, &c., and the entire roof, as well as, all the guttering, rain pipes, &c., of the New Houses of Parliament, are composed of Galvanized Sheet Iron.

Glebe has the largest number of 1860s-1870s cottages and terraces forming one continuous townscape in Australia. The pre-rehabilitation photographs reveal the deplorable state of many of these residences before the 1970s. One of the difficulties in the early days of the restoration work was to find qualified tradesmen such as slaters and workers familiar with the galvanised iron work to make the veranda canopies. The few older workers who had these skills passed their knowledge onto younger men during reconstruction. Improved materials were used in some instances. Laminated marine ply was used for barge boards because it weathers better than the original solid timber.

Note the attics in the roofs of nos 84-82 and no. 90. The latter is Thorpe House built by Richard Gawthorpe in 1867. He lived here until 1891. It has an iron canopied veranda and the attic is in the centre of the sloping roof. No. 105 was built in 1869 but has 1920s bungalow additions. Nos 104-106 are post-Regency semi-detached cottages. Note the sturdy twin wooden veranda posts, shutters, and three chimneys. It is believed architect Edmund Blacket built these houses.

On the right note nos 107-109. There are round-headed lights in the upper sashes of the windows, popular in Glebe in the late 1870s - the houses date from 1876. Nos 111-113 were built in 1869 and are the oldest two-storeyed residences in this area.

Nos 115-117 are set back from the street so there are small gardens. Note the original wooden frieze, fretted brackets, and turned veranda posts. These two houses were built by a Mr Gracie before 1870 in a post-Regency design. No. 117 has a memorial plaque to the late Jessie Guthrie. The builder, Mr Gracie, was her grandfather; and Jessie Guthrie lived at no. 117. The plaque commemorates the role she played in setting up Elsie, Australia's first women's refuge in 1973. The original Elsie was located in Westmoreland Street.

You are nearly back to Record Reign Hall. Turn right and return to Glebe Point Road for refreshments.

Profiles

DR WILLIAM BALMAIN 1762-1803

William Balmain was born to a tenant farmer, Alexander Balmain, and his second wife, Jean Henderson, in Balhepburn, Perthshire, Scotland on 2 February 1762. A medical student at Edinburgh University, when William was eighteen he entered the Royal Navy as a Surgeon's Mate. When the First Fleet assembled to colonise New South Wales, Balmain was assigned to the transport *Alexander*.

Dr Balmain probably saved the life of the founding governor of the colony, Arthur Phillip. Phillip was speared at Manly on 7 September 1790 by a frightened Aborigine and the governor was hastily rowed to Sydney where Balmain removed the spear 'with great skill'.

The doctor spent four years at Norfolk Island from 1791 until 1795 when he returned to Sydney as Principal Surgeon of the colony, succeeding Surgeon White. In 1788 Balmain had fought a duel with White. Balmain had acted as Civil Magistrate on Norfolk Island and fulfilled this role also in Sydney where he became embroiled in an argument with the fiery John Macarthur. Balmain was actively involved in trade in the colony, especially in importing rum. He was also the first Collector of Customs under Governor King in 1800.

Numerous lands grants were given to Balmain including, in 1800, 550 acres (222.5 ha.) in today's Balmain. This grant was later transferred to John Gilchrist.

Balmain returned in ill health to England in 1801 and died on 17 November 1803. He had a son and two daughters born in New South Wales by a convict woman, Margaret Dawson. His son also became a surgeon, later practising in Sydney. His eldest daughter died in Sydney at three years of age. A fourth daughter was born on the day Balmain was buried, 25 November 1803, but appears to have died in infancy. The children were baptised in the name of Henderson.

JOSEPH LOOKE 1803-1868

Joseph Looke was born in England to a mail-coach driver and his wife in 1803. Joseph emigrated to Australia in 1832 with his wife, Hannah, two sons and a daughter. He came under the bounty system which encouraged tradesmen to settle in New South Wales.

Looke established a boatbuilding firm at Darling Harbour and by 1838 had purchased land in Balmain. By 1844 he owned a boat yard, wharf, timberyard and had built his growing family of eight children a fine stone cottage. He owned other Balmain cottages. Looke expanded his business into the coal trade and established a coalyard on his Balmain waterfront — steamers needed bunkering in Sydney. He continued to prosper and built more houses — Braeside (by 1850) and Cliffdale House (by 1866). Looke's eldest daughter married Captain John McKinlay and resided at Braeside. Looke in 1866 began building terrace houses including Alfred Terrace in Looke's Avenue, which consisted of three brick and stucco houses. Looke apparently intended to add to the terraces but died suddenly in 1868. He was found floating in the harbour on 30 May close to his wharf. His wife had predeceased him in 1867.

His business was continued by his son, William.

WILLIAM EDMUND KEMP 1831-1898

William Kemp was born at Stroud, New South Wales, to Simon and Mary Ann Kemp on 9 January 1831. In 1849 when he was eighteen he became a pupil of architect Edmund Blacket and later joined the office of the Colonial Architect, William Weaver.

In 1857 Kemp joined Weaver in private practice and was engaged in rebuilding St Mary's Church, Balmain. In 1872 Kemp rejoined the Colonial Architect's Office under James Barnet. In 1880 William Kemp was appointed Architect for Public Instruction and began to produce school buildings of secular architecture, adapting the Classical architectural of the Italian Renaissance.

In 1876 he drew up plans for the Nicholson Street School at Balmain. Building began in 1882. Kemp built many New South Wales schools including the Pyrmont School at John and Mount streets in 1891. His restrained design at this school is characteristic of the revolt against Victorian ornamentation. Architect Morton Herman said in his work *The Architecture of Victorian Sydney*

> They (the architects) began to feel the form, the whole mass of the building, should be more important than the parts.

Kemp also designed new buildings at Ultimo for the Sydney Technical College, which was completed in 1891. He designed the Technological Museum as part of the college buildings in 1892.

William Edmund Kemp was brother to Charles, a proprietor of the *Sydney Herald* (later *Sydney Morning Herald*) with John Fairfax.

Kemp retired in 1896 after suffering from a long and severe illness. He died at his home at Stanmore on 14 June 1898.

WILLIAM MORRIS HUGHES 1864-1952

William Morris Hughes was born in London to Welsh parents on 25 September 1864. He was educated in Wales and at St Stephen's School, Westminster. He emigrated to Australia in 1884 and arrived in Brisbane and worked in many jobs before coming to Sydney in 1886. He again worked in various positions and became interested in the labour movement. Eventually he became president of the Waterside Workers' Federation, was elected to the New South Wales Parliament in 1894, and became a skilful debator. He supported an eight-hour working day and became an advocate of Federation. In 1901 he was elected to the federal House of Representatives and later served as Minister for External Affairs and Attorney-General. In 1915 William Hughes became Prime Minister. From 1901 Hughes had advocated compulsory military training, but in 1916 the majority of the Labor Party was opposed to conscription and the issue split the Party. In that year, a referendum of the Australian people rejected conscription. Hughes and twenty-three supporters left the Labor Party and joined with the conservatives in 1917 to form a National Party government of which Hughes became Prime Minister.

Hughes attended the Peace Conference at Versailles and supported the demand for heavy reparations from Germany. Country Party pressure in 1923 culminated in Hughes's resignation as Prime Minister on 9 February 1923. In his later years in politics he attempted to form a new party, the Australia Party, but failed and joined the United Australia Party led by J. A. Lyons. Hughes became Minister for Repatriation and Health in 1934 and began his 'populate or perish' campaign to counteract Australia's falling birth rate.

In 1939 R. G. Menzies defeated Hughes in a contest for the prime ministership. Hughes became Minister for the Navy and Attorney-General until 1941 when Labor won office under John Curtin. In 1945 Hughes joined the newly created Liberal Party and remained a member until his death in 1952.

He was an author, speaker, and dominant public figure. In 1919 he had been a hero to servicemen and as 'the little digger' was always a familiar figure at Anzac Day marches. He refused to delegate authority and was known to be a difficult man. His securing of the mandate over German New Guinea was probably his most significant political achievement.

CAPTAIN THOMAS STEPHENSON ROWNTREE 1818-1902

Thomas Stephenson Rowntree was born at Sunderland, County Durham, England, on 6 July 1818. In 1838 when he was twenty he went to sea as a shipwright but within four years held his Master's Ticket and was trading in the Mediterranean and Baltic.

When gold was discovered in the Australian colonies Rowntree and a partner built the *Lizzie Webber* of 206 tons and sailed with 100 passengers for Australia. In Melbourne the vessel proved too large for the wharf and Rowntree sailed to Sydney and anchored in Waterview Bay.

He settled in Balmain in 1851 and in 1853 planned a patent slip on Waterview Bay. Thomas Sutcliffe Mort persuaded Rowntree to build a dry dock and Mort, merchant J. S. Mitchell, and Rowntree formed Rowntree & Co. in 1854. Rowntree later left the business and began building composite screw steamers. He then sailed for New Zealand on a saw-milling venture but returned to Sydney in 1869.

Captain Rowntree was a foundation alderman of the Municipality of Balmain and was mayor for two terms. He built Northumberland House at 87 Darling Street and Woodleigh, designed by Edmund Blacket, in Stack Street for his son, Thomas.

Captain Rowntree died at Northumberland House on 17 December 1902 when he was eighty-five and was buried in the old Balmain Cemetery. The stone pedestal from his grave is now a monument to Captain Rowntree in Macquarie Terrace, Balmain.

GEORGE ALLEN 1800-1877

George Allen was born in November 1800 to Dr Richard Allen and his second wife, Mary Lickfold. His father was a favourite physician to the Prince Regent but died when George was only six years old.

Dr Allen had operated a business selling bottles of medicine made to his prescription to the public. The business was managed by a Thomas Collicott. On the death of Dr Allen, Collicott retained the business and later married Mrs Allen.

In 1812 because of the Napoleonic wars, the British Government kept a watchful eye on tax evaders. When Collicott omitted to affix revenue stamps to his medicine bottles, he was apprehended by the authorities and sentenced to be transported. Mrs Collicott sought the assistance of Sir Robert Wigram, a friend of her late husband. Sir Robert secured passages for her and six children on the *Mary Anne* and also provided a letter of introduction to Governor Macquarie. The vessel sailed on

26 July 1815 with young George Allen who was aged fifteen.

On arrival in the colony Collicott's eldest son received a grant of 200 acres (81 ha.) and Governor Macquarie informed George Allen he would receive a similar grant when he attained eighteen years. In the meantime Macquarie arranged for George to be articled to the Crown solicitor, W. H. Moore. When Moore ran into conflict with the governor, George was forced to forego his legal training and work as a clerk to the Government Printer, George Howe.

He later completed his legal training under Frederick Garling and on 22 July 1822 was admitted to practice. George Allen was the first solicitor to serve his five years of articles in the colony.

In 1823 he married Jane Bowden, daughter of the schoolmaster of the Male Orphan School. He became a leading member of the Methodist Society and by the 1840s had a thriving legal practice in Elizabeth Street. Allen also became chairman of the Australian Gas Light Company and served on the Council of Education.

His last years were marred by blindness. He died on 3 November 1877 in his seventy-seventh year.

GEORGE ALLEN MANSFIELD 1834-1908

George Allen Mansfield was born in 1834, the son of the Reverend Ralph Mansfield. The Reverend Mansfield had arrived in the colony in 1820 as a Wesleyan minister and served at Parramatta, Windsor, and Hobart. He left the ministry in 1828 and became editor of the *Sydney Gazette* from 1829 to 1832.

George Allen Mansfield was articled to John Frederick Hilly, architect of Sydney's Royal Exchange, in 1850 but soon set up a private practice. In 1857 Mansfield was appointed Architect of Public Schools — his schools included Crown Street, Cleveland Street, and Surry Hills. In 1860 he was awarded third place in a competition to design of new Houses of Parliament in Sydney. He was offered the position of Colonial Architect but declined. Mansfield was a founder of the Institute of Architects and elected its first president in 1871. His bust is in the headquarters of the Royal Australian Institute of Architects. He was the first Australian to be elected a Fellow of the Royal Institute of British Architects in 1873. With architects Blacket and Blackman, Mansfield reported on the defective foundations of Sydney Town Hall.

Mansfield was prolific and designed the Royal Prince Alfred Hospital, the AMP building in Pitt Street, Savings Bank of New South Wales, City Mutual Life Assurance Company's offices, and over forty banks for the Commercial Banking Company of Sydney Limited. Mansfield also designed the Hotel Australia, visiting the United States of America to study hotels before completing his design.

Mansfield was active in the movement to revive a Volunteer Defence movement and enrolled in the Glebe Company, holding a commission as First Lieutenant. He also served as an alderman in Glebe.

George Allen Mansfield married Emma, daughter of Sir George Wigram Allen. He once argued with the Methodist minister of Glebe and fought a duel on the parsonage lawn. He subsequently became an Anglican.

George Mansfield retired from his practice in 1905 and died in January 1908. He left his widow, Emma, and seven children.

Time Line

(A chronicle of significant dates in world history 1606–1901)

1606   Torres sailed through Torres Strait.
        Dutch vessel Duyfken discovered Cape York Peninsula.
1613   Romanov dynasty founded in Russia.
1616   Death of Shakespeare.
        Dirk Hartog landed off Western Australian coast.
1620   Pilgrim Fathers landed in New England.
1625   Charles I King of England.
1642   Tasman discovered Van Diemen's Land.
1648   Cromwell defeated Royalists.
1649   Charles I executed.
1660   Charles II King of England.
1662   Royal Society founded in London.
1688   William Dampier observed Western Australia.
        British Parliament passed Bill of Rights.
1692   Massacre of Glencoe.
1699   Dampier visited New Holland.
1714   Queen Anne died — George I King of England.
1725   Peter the Great of Russia died.
1726   Swift's *Gulliver's Travels* published.
        Isaac Newton died.
1750   Johann Sebastian Bach died.
1760   George III King of England.
1769   James Watt perfected steam engine.
1770   Captain Cook discovered east coast of Australia.
1774   Louis XVI King of France.
1779   Captain Cook killed in Hawaiian Islands.
1788   First Fleet arrived Botany Bay.
1788   26 January settlement at Sydney Cove. Population of New South Wales approximately 1,030.

1789	Bastille stormed in Paris.
	George Washington first President, United States of America.
1790	John Macarthur arrived Sydney on Second Fleet.
1791	First land granted to James Ruse of thirty acres (12 ha.) at Parramatta.
	Wolfgang Amadeus Mozart died.
1793	Louis XVI of France guillotined 21 January.
	Marie Antoinette guillotined 16 October.
	First free settlers arrived in Sydney.
1794	Samuel Marsden arrived in the colony.
	End of Reign of Terror, French Revolution.
1795	John Hunter, Governor of New South Wales.
	Warren Hastings acquitted in England.
1798	First church in Sydney destroyed by fire.
	Napoleon in Egypt — Battle of the Nile.
	Irish rising.
1800	Philip Gidley King, Governor of New South Wales.
1801	First muster taken in New South Wales.
	Population 5,217.
	Legislative union of Ireland and England.
1803	*Sydney Gazette* founded by George Howe.
1804	Battle of Vinegar Hill.
	Napoleon crowned Emperor of France.
1805	Battle of Trafalgar.
1806	William Bligh, Governor of New South Wales.
	Death of William Pitt the younger.
1807	First export of wool from New South Wales.
	Dissolution of German empire.
1808	Rum Rebellion.
	Napoleon entered Rome.
1810	Lachlan Macquarie, Governor of New South Wales.
	Population of colonies 11,566.
	Isaac Nichols appointed first Postmaster in Sydney.
1811	General Hospital (Rum Hospital) foundation stone laid.
	Luddite riots in England.
1812	Floods on Hawkesbury River — rose twelve feet (3.75 m.).
	First hops gathered in colony.
	Napoleon's retreat from Moscow.
1813	Blue Mountains crossed by Blaxland, Lawson, and Wentworth.
1814	Francis Greenway arrived in Sydney.
	Napoleon abdicated.
1815	William Cox completed road over the Blue Mountains.
	Battle of Waterloo.
1817	Bank of New South Wales founded in Sydney.
	Riots in Manchester.
1819	Commissioner Bigge arrived in New South Wales.
	Peterloo massacre, Manchester.

1820	Death of Sir Joseph Banks in England.
	Death of George III.
	Population of colonies 33,543.
1821	Sir Thomas Brisbane, Governor of New South Wales.
	John Keats, poet, died.
1822	Governor Macquarie left New South Wales.
	Percy Shelley, poet, died.
1823	New South Wales Judicature Act passed — first step towards representative government.
1824	Death of Lachlan Macquarie in London.
	Bolivar Dictator of Peru.
1825	Van Diemen's Land became a separate colony.
	Sir Ralph Darling, Governor of New South Wales.
	First public railway opened between Stockton and Darlington, England.
1826	Sudds and Thompson Case in Sydney.
	Menai Suspension Bridge opened England/Wales.
1828	Russia declared war on Turkey.
	First regular census in colonies.
1829	Settlement at Swan River, Western Australia.
	First strike in Australia by newspaper compositors of *The Australian*.
1830	First civil jury sat in the Supreme Court.
	Population 70,039 in colonies.
1831	Sir Richard Bourke, Governor of New South Wales.
	Epidemic of cholera in England.
1832	King's School, Parramatta opened.
	Goethe, German poet, died.
1833	British Parliament passed Act abolishing slavery in British colonies.
1834	John Macarthur died at Camden.
	Faraday discovered electric self-induction.
1835	Death of Elizabeth Macquarie in Scotland.
	Mark Twain born.
1836	Settlement of South Australia.
	Australian Museum founded.
	First telegraph invented by Samuel Morse.
1837	Melbourne laid out.
	Architect Francis Greenway died — Drought lasts until 1839.
	Victoria, Queen of England.
1838	Sir George Gipps, Governor of New South Wales.
	Coronation of Queen Victoria.
1840	Cessation of convict transportation.
	Population of colonies 190,408.
	Queen Victoria married Prince Albert of Saxe-Coburg-Gotha.
1842	Last issue of the *Sydney Gazette*.
	Chartist procession in London.
1843	Bank of Australia closed its doors — 1840s depression.
	New South Wales granted representative government.
	Irish patriot Daniel O'Connell arrested in Ireland.

1846	Sir Charles Fitzroy, Governor of New South Wales.
	Irish potato famine.
	Corn Laws repealed.
1847	Lady Fitzroy killed at Parramatta.
	Arctic explorer Sir John Franklin discovered northwest passage.
	10-hour Factory Act passed in England.
1849	Convict ship *Hashemy* arrived in Sydney causing anti-transportation demonstrations.
1850	English poet, William Wordsworth, died.
	Population in colonies 405,356.
1851	Victoria became a separate colony.
	Discovery of gold by Edward Hargraves in New South Wales.
	Great Exhibition in London.
1854	James Barnet, architect, arrived in Sydney.
	Charge of the Light Brigade — Crimean War.
1855	Sir William Denison, Governor of New South Wales.
	Royal Mint opened in Sydney — Van Diemen's Land became Tasmania — First rail line between Parramatta and Sydney.
	Great Exhibition in Paris.
1857	Wreck of the *Dunbar*.
	Indian Mutiny.
1858	Australian population passsed 1,000,000.
	Siege of Lucknow, India.
1859	Queensland became a separate colony.
	Darwin's theory of evolution published.
1860	Anti-Chinese riots at Lambing Flat.
	Australian population 1,145,585.
	Abraham Lincoln, President, USA.
1861	Victor Emmanuel first King of Italy.
	Civil War, USA.
1863	Northern Territory established.
1865	St Mary's Cathedral destroyed by fire in Sydney.
	President Lincoln assassinated.
1867	Henry Lawson born near Grenfell, New South Wales.
	Emperor Maximilian of Mexico shot.
1868	Duke of Edinburgh shot at Clontarf, near Sydney.
1870	Novelist Charles Dickens died.
	Australian population 1,647,756.
1872	Henry Parkes forms first Ministry in New South Wales.
	The Ballot introduced in England.
1879	Stock Tailoresses' Union formed.
1877	Australian population exceeds 2,000,000.
	Victoria, Empress of India.
1879	International Exhibition held in Sydney.
	Garden Palace built.
	First successful cargo of frozen meat sent to England.

Edison invented incandescent filament lamp.
Transvaal declared a republic.
1880 Henry Parkes introduced Public Instruction Act giving a free, compulsory, and secular education.
1882 Charles Darwin died. Tailoresses' strike — December to January 1883.
1883 Karl Marx died.
1884 Smallpox epidemic — first Australian public health conference held.
1885 General Gordon killed at Khartoum in the Sudan.
Sudan Contingent sailed from Sydney.
1886 Gladstone's first Irish Home Rule Bill.
1888 Drought — New South Wales has driest year since records began.
1889 Tenterfield speech on Federation by Henry Parkes.
1890 Bismarck dismissed in Germany.
1891 Australian population 3,177,823.
1892 Financial crisis — 1890s depression.
1896 Sir Henry Parkes died.
1898 David Scott Mitchell disclosed he intended to bequeath his library of Australiana to the State.
Pierre and Marie Curie isolated polonium and radium.
1899 Boer War began.
1900 Delegates from Australia visit London when Commonwealth Bill submitted to Imperial Parliament.
Boxer risings in China.
Freud published *Interpretation of Dreams*.
1901 Commonwealth of Australia proclaimed.
Australian population 3,773,801.
Queen Victoria died.

Opening Times

Because of the risk of vandalism most churches are not permanently open.

The Watch House, Balmain: open Saturdays 1.00 pm to 3.00 pm.

Lyndhurst 61 Darghan Street, Glebe: Telephone 692 8366. Lyndhurst is the headquarters of the Historic Houses Trust of New South Wales. The ground floor is used as a conservation resource centre for historic houses. It is advisable to telephone if you wish to make use of this resource.

Further Reading

Peter L. Reynolds and Paul V. Flottmann *Half a Thousand Acres — Balmain — A History of the Land Grant* Sydney: The Balmain Association, 1976.
H. V. Evatt *Rum Rebellion* Sydney: Angus & Robertson, 1938.
Ruth Park *The Companion Guide to Sydney* Sydney: Collins, 1973.
Cedric Emanuel and Olaf Ruhen *Balmain Sketchbook* Sydney: Rigby, 1975.
Simon Davies *The Islands of Sydney Harbour* Sydney: Hale & Iremonger, 1984.
Peter Reynolds and Robert Irving *Balmain in Time* Sydney: The Balmain Association, 1971.
Debbie Nicholls *Around Balmain* Sydney: The Balmain Association, 1986.

Index

Abbey, The (Annandale) 77
Abbey, The (Glebe) 94
Aboriginal Training & Cultural Institute 37
Abrahams, Esther 59, 61
Ada View 21
Adam, Captain William 46
Alderly, Balmain 45
Allen, George 86, 88
Allen, George Wigram 86, 87, 89, 91
Allen, Dr Richard 86
Anchorage, The 46, 50
Annandale Association 71, 74
Annandale Borough Council 71
Annandale House 60, 63
Annandale Public School 63
Annandale Theatre 77
Aqueduct 78
Arthurleigh, Annandale 63
Australian Labor Party 26

Ball, Lieut. Henry L. 12
Balmain Almanac 26, 36, 46, 47
Balmain Association 29, 31
Balmain Bowling Club 32
Balmain Cemetery 57
Balmain coalmine 29, 49
Balmain Fire Station 44
Balmain Hospital 45
Balmain, John 17
Balmain Municipal Council 43, 44, 56
Balmain Regatta 11, 12, 55
Balmain Town Hall 43
Balmain village 32

Balmain Watch House 29, 31, 32
Balmain, Dr William 11, 17
Bank of New South Wales 39, 50, 64
Barnard, John Vinson 17
Barnet, James 17, 44, 47, 89
Barton, Sir Edmund 67
Bates, E. A. 39
Beale's Piano Factory, Annandale 67
Beattie, James 16
Bell, John 15
Bibb, John 35
Bidura, Glebe 45
Birch Grove House 45, 51, 54
Birch, John 45, 52
Birchgrove Park and Oval 51
Birchgrove Public School 48, 49
Birchgrove village 56
Bishopthorpe, Glebe 83, 95, 96
Blacket, Arthur & Cyril 65, 95
Blacket, Edmund 17, 24, 26, 31, 37, 44, 45, 63, 65, 89, 95, 100, 101
Blacket, Sarah 45
Blackman, Charles 74, 77
Blake, Robert 33, 35
Bligh, Governor 60
Bondi outfall 78
Bowden, Jane 86
Bowman, Dr James 95
Boyce, Rev. W. B. 91
Bradman, Don 27
Branksea 19
Broomfield, Captain John 24
Broughton, Bishop 83, 96
Browne, Robert 41

Buchanan, E. H. 43, 50
Bull, William 92
Bullivant, Charles 16
Burt, James and William 41

Cahermore, Balmain 16
Cakebread, Rev. W. T. 50
Cameron, Charles C. 33
Cameron, Ewen 56
Cameron, Major Ewen Wallace 33
Cameron, John 35
Cameron's Cove 33
Canterbury Vale 83
Cavill, John 15, 24
Chape, Alexander 40
Chape, Catherine 40
Charles' Villas 23
Charlton, Rev. W. A. 50
Chinese Joss House, Glebe 84
Cliffdale House 46
Clifton Villa 56
Clune, Frank 93
Cockatoo Island 21, 23, 47, 48
Colgate-Palmolive factory 29
Collicott, Thomas 86
Convent of the Immaculate Conception 37
Cormack, Alexander William 52
Cornell, John 24
Council Chambers, Annandale 71
Cremorne 49

Darby, Cecil West 78
Darling Harbour 21
Darling, Sir Ralph 14
Deloitte, Captain William 54
Deloitte, Quartin 55
Demeter's Bakery, Glebe 99, 100
Dennis, William 21
Desmond Villa 21
Dobbie, John 15
Downes, Mrs. Elizabeth 84

Eastcliffe 19
Edwinville, Annandale 68
Elizabeth Terrace 24
Elizabeth's Villas 23
Elkington Park 46, 47
Evatt, Dr H. V. 24
Ewenton, Balmain 33

Fenwick, J. & Co. Pty Ltd 12, 15
FitzRoy, Governor Sir Charles 28, 47
Ford, Robert 41
Forsyth, Commander Leonard 47

Foss, Ambrose 94
Fraser, Dawn 46
Fulloon, Elizabeth 52

Garden Palace 89
Gardner family 23
Gawthorpe, Richard 101
Geierstein, Birchgrove 52
Gibson, John 51
Gilchrist Educational Trust 12
Gilchrist, John 11, 14, 15
Gipps, Governor 47
Gladstone Park Reservoir 39
Gladstone Park 15, 17, 39
Glebe Estate 96, 100
Glebe Island 33
Glebe Society 96, 97
Glenree 51
Glentworth 24
Goat Island 12
Good Samaritan Convent 87
Goodman, Walter 61
Goodman's Building, Annandale 61, 62
Gordon, Rev. Thomas 32
Grange, The 36
Greenway, Francis 61
Guthrie, Jessie 101

Hackland Carrying Company, Glebe 98
Hackland, John 98
Hallen, Edward 92, 94
Hammond, Canon R. B. S. 92
Hampton Villa 35
Harbour View Terrace 21
Harold Park Racecourse 86, 88
Hartford, Glebe 83
Hereford House, Glebe 92, 94, 95
Hermitage, The (Glebe) 94
Hilly, J. F. 37
Hinsby Reserve 73
Historic Houses Trust 95
Hixson, Captain Francis 19
Hoad, Lew 92
Hope, Rev. John 91
Hopson, Nicholson 21
Hotels
    Albion 40
    Dolphin 15
    London 40
    Marquis Arms 16
    Marquis of Waterford 16
    New Unity Hall 19
    Riverview 46
    Rob Roy 36

Index

Shipwright's Arms 15
Unity Hall 43
William Wallace 56
Howard Smith site 49, 50
Hughes, Prime Minister 'Billy' 24, 43
Hunt, Hon. Edward 35
Hunt, J. Horbury 37, 95
Hunter Baillie, John 64
Hunter Baillie Memorial Church 64-68, 71
Hunter, Governor 59, 60

Illoura Reserve 20
Iron Cove Bridge 48

Jaques, Major 19
Johnson, Rev. Richard 81
Johnston, Captain George 33, 59, 60, 61. 63
Johnston, Robert 61, 71, 74, 78
Johnston's Bay 33
Johnston's Creek 78
Jones, Alderman Evan 84
Joubert, Didier Numa 50

Keda, Birchgrove 51
Kelly, Archbishop 66
Kemp, William Edmund 17, 26, 31
Kentville, Annandale 74
Kerr, Sir John 24
Kinchela, Attorney General John 94
King, Copeland 50
King, Philip Gidley 24
King, Rev. Raymond 50
Kinwarra 35
Kirkpatrick, James 92, 93
Kirribee, Glebe 92

Lang, Dr John Dunmore 32, 64
Lappan, Adam 72
Lawrence, George 19
Leeds, Joseph 92
Leichhardt Municipal Council 27, 44, 71, 96
Lemm, Frederick 55
Lenardville, Birchgrove 51
Lewis, George 62
Lindsay, Norman 49
Loane, R. W. 45
Lodge, The (Glebe) 89
Logan Brae 50
Long Nose Point 52
Looke, Joseph 16
Lord, John 51

Lyndhurst, Glebe 95
Lyons, Captain John 19

Machon, Mary 98
Mackenzie, Rev. Peter Falconer 65
Mackie, Helen Hay 64
Macquarie, Governor Lachlan 37, 40, 45, 60, 86
Macquarie Lodge 62
Mansfield, George Allen 68, 87, 89
McDonald, James 32, 33, 40, 41, 43
McKenzie, Bruce 20, 22
McKenzie, Henry 15
McLean, Captain 14
McRae, George 96
Melba, Dame Nellie 27
Minton tiles 67
Mitchell, J. S. 28
Moorfield 35
Morrison & Sinclair 52
Mort Bay 26, 27 28
Mort, Thomas Sutcliffe 26, 27, 33, 56, 89
Mort's Dock 21, 27, 28, 29
Murcutt, Glen 21

Nichols, Isaac 60
Nicholson Street Public School 17
North Annandale Public School 73
North Sydney 31
Norton House, Annandale 64

Oddfellows Hall 40
Onkaparinga 19
O'Ridge family 63
Orphan School Creek 92
Oybin, Annandale 74

Parkes, Sir Henry 24, 35, 54, 74
Peacock, John Thomas 20
Peacock Point 12, 14, 20
Pearson, Captain 19
Perkins, J. 56
Phar Lap 27
Phillip, Governor 81
Pigeon Ground School 39
Pinetree House 26
Plym Terrace 15
Polding, Archbishop 37
Pyrmont 21, 33, 35

Ravens Court, Birchgrove 51
Raywell, Birchgrove 52
Rembold, Charles 23
Reuss, Ferdinand, Jnr 62

Reuss, Ferdinand, Snr 94
Richards, John 74
Roach, Captain John 46
Roberts, Tom 27
Rose family 23
Row, Edward 17
Row, John 17
Rowe, Thomas 89, 92, 94
Rowntree, Captain Thomas Stephenson 27, 28, 43, 57
Rowntree Monument 56
Royal Exchange, Sydney 37
Rum Rebellion 60
Ruse, James 24
Russell, John Peter 14
Russell, Peter Nichol 14

Shannon Grove 35
Sheerin, Joseph 62
Ships
   Clarkeston 54
   Herald 47
   Hunter 28
   Lady Penrhyn 59, 60
   Lizzie Webber 27, 28
   Prince George 46
   Sabraon 50
   Sixty Milers 33
   Tingira 50
Simonette, Achille 54
Sims, John 23
Smith, Francis 21
Snails Bay 52
Snapper Island 47
Snow, William Robert 19
Spectacle Island 47
St Aidan's Church, Annandale 66, 72
St Andrew's Congregational Church 41
St Augustine's Church, Balmain 29, 37, 39
St Barnabas' Church, Broadway 93
St Brendan's Church, Annandale 66
St James Church 93
St James Presbytery 93
St James Primary School 93
St John's Parish Hall, Glebe 96
St John's the Evangelist, Birchgrove 50
St John's the Evangelist, Glebe 95
St Joseph's Convent 67
St Mary's Cathedral, Sydney 37, 74, 77
St Mary's Church, Balmain 26
St Philip's, Church Hill, 83, 96
St Scholastica's College, Glebe 86, 87
Stainger, Mrs 33
Steiner, Rudolph 100

Stone, Louis 24
Sturdee, E. J. 50
Suddy, James 21
Sulman, John 74, 77
Swan, Henry B. 33
Swan, L. B. 33
Sze Yup Temple 84

Taylor, Sir Allen 67
Therry, Rev. John Joseph 37
Thornton Park 14
Thorpe House, Glebe 101
Throsby, Charles 36
Throsby, Jane 36
Thunderbolt, Captain 23
Toxteth House, Glebe 73, 78, 86, 87, 88, 89
Tramways 14, 15
Tranby, Glebe 91
Trickett, Ted 55
Trouton, Captain Frederick 46
Tugs 12
Turner, George 16

Uniting Church, Annandale 71
Unity Square 43
University of Sydney 14, 44, 63, 74, 77, 89

Verey, George 40
Verge, John 35, 54, 86, 94
Vernon, Walter Liberty 97
Vidette, Birchgrove 50
Vinegar Hill 60

Waterman's Cottage 15
Waterview Bay 26, 27, 43
Waterview Dock 28
Weaver, William 26
Wells, Rachel 52
White Bay 21, 33
White, John 21
Whitfield, George 45
Whitlam, Gough 97
Wigram, Sir Robert 86
Wilkinson, Judge 94
Williams, George 92
Witches' Houses, Annandale 74
Woolley, Thomas 92, 93, 94
Woolwich 29
Wootten, Annandale 67
Workingmen's Institute 40
Wran, Neville 17, 19
Wright, John William 15
Wynthorpe, Glebe 98

Young John, 71, 74, 77, 89
Young, William Adolphus 36, 37